Klaus Baumgärtel
Konrad Sauer

Topics on Nonlinear Wave-Plasma Interaction

1987

Birkhäuser Verlag
Basel · Boston · Stuttgart

Authors
Klaus Baumgärtel
Konrad Sauer
Institut für Kosmosforschung
Akademie der Wissenschaften der DDR, Berlin

Reproduction of the original authors' manuscript

CIP-Kurztitelaufnahme der Deutschen Bibliothek

Baumgärtel, Klaus:
Topics on nonlinear wave plasma interaction /
Klaus Baumgärtel ; Konrad Sauer. — Basel ; Boston ;
Stuttgart : Birkhäuser, 1987.
ISBN 3-7643-1864-3 (Basel ...)
ISBN 0-8176-1864-3 (Boston)
NE: Sauer, Konrad:

Lizenzausgabe für alle nichtsozialistischen Länder:
Birkhäuser Verlag, Basel 1987
Printed in GDR
ISBN 3-7643-1864-3
ISBN 0-8176-1864-3

Preface

The text presented here is an extended english version of a report by the authors which appeared in April 1983 at the Institute of Cosmical Research of the Academy of Sciences of the GDR in German. It covers several selected topics on nonlinear wave-plasma interaction in a treatment based on a hydrodynamic plasma description. Thus, no attempt has been made to give a comprehensive view on all aspects of the interaction of strong electromagnetic waves with plasmas. The text is partly introductory and presents partly current results. The authors hope that it will be of interest to students and scientists not only in the field of plasma physics.

The authors thank Akademie-Verlag, Berlin and Birkhäuser-Verlag, Basel for their encouragement to prepare the English manuscript and Mrs. Ch. Geier for carefully typing the final off-set version.

Klaus Baumgärtel
Konrad Sauer

Berlin, in April 1986

Contents

Introduction

This monograph is devoted to theoretical research on the interaction of intense electromagnetic radiation with unmagnetized fully ionized plasmas. It derives partly from a number of theoretical numerical investigations of the authors on processes as harmonic generation, stimulated Mandelstam-Brillouin scattering, stimulated Raman scattering and density profile modification. In all of these cases the plasma reacts nonlinearly on an externally applied electromagnetic wave.

The study of nonlinear effects at electromagnetic wave propagation in plasmas was initiated in the thirties by the discovery of nonlinear interaction of radio waves in the ionosphere (cross modulation, self-interaction, generation of combination frequencies). The underlying basic nonlinearity is connected with the collisional heating of electrons in the electric field of the wave, typical for weakly ionized collision-dominated plasmas. Heating of the electrons leads, first, to a change of the collision rates of electron-ion and electron-neutral collisions and thus changes the absorption. This, in turn, produces an essentially nonlinear dependence of the field amplitude in the interior of the plasma on the field amplitude of the wave incident on the plasma boundary. Heating of the electrons may also change the refractive index of the wave via a change of the plasma density due to a disturbance of the ionization balance and/or enhanced electron pressure. (For relevant literature see General References.)

Due to the rapid increase in power and directivity of radio wave transmitters these processes are still of current interest. A recent aspect in this field is the artificial modification of the properties of the ionosphere by applying radio waves.

The interest in the interaction of intense electromagnetic radiation with fully ionized plasmas arose in connection with the efforts to bring about laser-driven fusion. When a high-power laser beam is focused onto a solid pellet a very short-lived, high-temperature plasma is formed at the surface. In order to create conditions for thermonuclear reactions in the interior of the pellet an effective deposition of the laser energy to thermal energy of the pellet via laser-plasma coupling is necessary. When laser light above certain power levels irradiates a plasma, collective nonlinear processes occur which can either enhance or retard the light absorption. Since energy absorption is crucial in laser fusion, a knowledge of the penetration depth allowed by these processes and of the reflected fraction of the incident energy is required.

Some of these nonlinear collective processes are well-known from nonlinear optics and the same denotations are used for their analogues in plasmas, e. g., stimulated Mandelstam-Brillouin scattering, stimulated Raman scattering. Thereby Langmuir waves and ion acoustic waves play the role of the optical and the acoustic phonons, respectively.

A plasma, however, exhibits some peculiarities compared with an optical medium which cause qualitatively new types of nonlinear interaction. Of particular importance is the existence of freely moving charged particles. Electrons or ions can be, for example, accelerated to very high energies in strongly localized electric fields. On the other hand, coherent interaction of a group of particles moving with the phase velocity of a wave provides an efficient mechanism for energy exchange between wave and plasma. These processes are associated with changes in the particle velocity distribution function and may cause a complex chain of nonlinear interactions.

Another important nonlinear mechanism operating in a collisionless plasma is connected to one of its macroscopic properties, namely, its compressibility. An inhomogeneous high-frequency field excerts a pressure on the plasma (ponderomotive force) and may produce strong distortions of the plasma density. This, in turn, changes the dielectric constant of the plasma which then becomes dependent on the electric field amplitude. This causes a nonlinearity of electrodynamic processes.

At optical frequencies, the power required to produce observable nonlinear effects in plasmas is large. At microwave frequencies, however, the magnitude of the nonlinear terms is much greater and therefore, it is not surprising that experimental observation of nonlinear effects in fully ionized plasmas was first accomplished at interaction of microwaves with large-volume laboratory plasmas. This was supported by the fact, that plasma diagnostics at microwave frequencies is considerably easier to handle than for optical frequencies.

Theoretical efforts to achieve a better understanding of the various nonlinear processes in a fully ionized plasma have been given at different levels according to the different mathematical models for the plasma description. The principal models are: (a) the hydrodynamic plasma model, (b) the kinetic plasma description and (c) particle simulation. All of these three codes have contributed to the progress in theory, partly in mutual control and interaction.

Whether one uses the fluid model, the kinetic description or particle simulation codes depends to a large extent on the problem one is considering. The hydrodynamic plasma model uses macroscopic quantities as density, flow velocity and thus fails in describing interaction processes with particles which result in a change of the particle velocity distribution function. The investigation of such features falls in the competence of kinetic theory or particle simulation codes. The fluid model, on the other hand, has the advantage of being mathematically considerably simpler. This makes it possible to study problems of practical interest involving spatial inhomogeneities and/or plasma flow.

Particle simulations can be considered as a kind of numerical experiments which allow in principle very general statements, the results of which are, however, often as complex as that of experiments and therefore call itself for interpretation by simpler models.

The present monograph is based on the hydrodynamic plasma description. The crucial problem is then the solution of the coupled system of hydrodynamic plasma equations and Maxwell's equations. For most of the problems a numerical approach is necessary. Numerical procedures have been used which are based on conventional discretization schemes of the underlying equations.

Throughout the book fully ionized plasmas with constant electron temperature are considered. Further, we restrict ourselves to a plane-layered medium; i. e., we assume that the properties of the plasma vary only in one space dimension. We shall not be concerned in this study with nonlinear processes produced by relativistic effects of particle motion at extremely high electric fields. Naturally, particular attention is paid to problems which the authors themselves contributed to.

The arrangement is as follows: The book is divided into two main parts dealing, respectively, with nonlinear processes where only electrons are involved (referred to as high-frequency processes) and phenomena which include ion motion (referred to as low-frequency processes).

As preliminary considerations Part I presents the derivation of the basic equations for both high-frequency and low-frequency processes from the two-fluid equations of a plasma and Part II summarizes basic features of linear wave propagation in nonuniform plasmas, as far as is necessary for the understanding of nonlinear processes. Thereby, the phenomenon of resonance absorption and the properties of the so-called structure resonances play a particular role by having independently attracted considerable interest.

Part III is devoted to high-frequency nonlinearities where the ions are considered to be fixed. It covers nonlinear phenomena as the generation of harmonics of the incident radiation frequency (Chapter 6), stimulated Raman scattering and two-plasmon decay (Chapter 7). The goal is to study these processes in nonuniform plasmas without making use of the WKB approximation. The calculations contribute to clarify the influence of the relevant physical parameters on the generation of the second and the third harmonic at a non-uniform plasma having a linear density profile. Moreover, results are presented concerning harmonic generation at plasma density profiles deviating from the linear one. As far as stimulated Raman scattering and two-plasmon decay is conderned, a hydrodynamic approach is naturally restricted to problems related to the initial weakly nonlinear stage of these processes. The strongly nonlinear stage and saturation mechanisms, which are intrinsically connected to fast electron generation, are beyond the scope of this treatise. Hence, the studies concentrate on the calculation of gradient-induced thresholds, of directly and indirectly generated scattering frequencies and the localization of the operation regimes. It turns out that the WKB approximation, even beyond its range of validity, yields results which do not differ significantly from the numerical calculations.

Part IV is devoted to low-frequency phenomena in which ion dynamics is involved. An extensive chapter is devoted to the theory of stimulated Brillouin scattering (Chapter 9); a phenomenon which is well established in the study of laser-plasma interaction. This contribution concentrates on the study of the space-time evolution of stimulated Brillouin scattering in a homogeneous plasma of finite length. Particular emphasis is given on effects produced, respectively, by (partially) reflecting boundaries (e. g., the influence of a feedback of the scattered wave on the development of the process) and by a plasma flow.

Chapters 10 and 11 are dealing with low-frequency interaction processes connected with strong density deformations in stationary and nonstationary treatment, respectively. Particular topics are the study of several kinds of strongly nonlinear stationary waves including envelope solitons, the anomalous nonlinear skin effect, the generation and the dynamics of density cavities, the demonstration of plasma acceleration by strong electromagnetic waves and the investigation of short-time modifications of the reflectivity of a nonuniform plasma (referred to as superreflection).

A list of General References is added, which contains several monographs and review papers more or less related to the topics of this treatise.

General References

Monographs

Akhmanov, S.A., Khokhlov, R.V.; Problems in Nonlinear Optics. Gordon and Breach, New York, 1972.

Blombergen, N.; Nonlinear Optics. W.A. Benjamin, New York, 1965.

Davidson, R.C.; Methods in Nonlinear Plasma Theory. Academic Press, New York and London, 1972.

Gekker, I.R.; Interaction of Intense Electromagnetic Fields with a Plasma (in Russian). Atomizdat, Moscow, 1978.

Ginzburg, V.L.; The Propagation of Electromagnetic Waves in Plasma. Pergamon, Oxford, 1964.

Gurevich, A.V., Shvartsburg, A.B.; Nonlinear Theory of Propagation of Radio Waves in the Ionosphere (in Russian). Nauka, Moscow, 1973.

Gurevich, A.V.; Nonlinear Phenomena in the Ionosphere. Springer-Verlag, Berlin, Heidelberg, New York, 1978.

Handbook of Plasma Physics, Vol. 1, 2. General Editors: M.N. Rosenbluth, R.Z. Sagdeev. North-Holland Publ. Comp., Amsterdam, New York, Oxford 1983.

Hora, H.; Nonlinear Plasma Dynamics at Laser Irradiation. Lecture Notes in Physics 102. Springer-Verlag, Berlin, Heidelberg, New York, 1979.

Litvak, A.G. (ed.); Interaction of Intense Electromagnetic Waves with a Collisionless Plasma (in Russian). Institute of Applied Physics IPFAN, Gorki, 1980.

Paul, H.; Nonlinear Optics (in German). Akademie-Verlag, Berlin, 1973.

Sagdeev, R.Z., Galeev, A.A.; Nonlinear Plasma Theory. W.A. Benjamin, New York, 1969.

Silin, V.P.; Parametric Interaction of High-Intensity Radiation with a Plasma (in Russian). Nauka, Moscow, 1973.

Weiland, J., Wilhelmsson, H.; Coherent Nonlinear Interaction of Waves in Plasmas. Pergamon, Oxford, 1977.

Rewiev Papers

Boyd, T.J.M.; Theory and Simulation of Electromagnetic Waves in Laser-Produced Plasmas. Lectures given at the 24. Scottish Universities Summer School in Physics, St. Andrews, Scotland, 1982.

Brueckner, K.A., Jorna, S.; Laser-driven Fusion. Rev. Mod. Phys. 46, 325(1974).

Chen, F.F.; Nonlinear Optics of Plasmas. Survey Lectures, Int. Conf. on Waves and Instabilities in Plasmas. Innsbruck, 1973.

Forslund, D.W., Goldstone, Ph.D.; Photon Impact - High-Energy Plasma Physics with CO_2 Lasers. Los Alamos Science Nr. 12, p. 2, Spring/Summer 1985.

Galeev, A.A., Sagdeev, R.Z.; Parametric Phenomena in a Plasma. Nucl. Fusion 13, 603(1973).

Ginzburg, V.L., Gurevich, A.V.; Nonlinear Effects in a Plasma Subject to an Alternating Electromagnetic Field (in German). Fortschr. d. Physik 8, 97(1960).

Goldman, M.V.; Strong Turbulence of Plasma Waves. Rev. Mod. Phys. 56, 709(1984).

Gorbunov, L.M.; Hydrodynamics of Plasmas in the Presence of a Strong High-Frequency Field (in Russian). Usp. Fiz. Nauk 109, 631(1973).

Gorbunov, L.M.; Interaction of Laser Radiation with Plasmas. Invited Lectures and Progress Report of SPIG 78. Dubrovnik, 1978.

Morales, G.J., Lee, Y.C.; Generation of density cavities and localized electric fields in a nonuniform plasma. Phys. Fluids 20, 1135(1977).

Kaup, D.J., Reiman, A., Bers, A.; Space-time evolution of nonlinear three-wave interactions I. Rev. Mod. Phys. 51, 275(1979).

Kennedy, E.T.; Plasma and Intense Laser Light. Contemp. Phys. 1984, Vol. 25, No.1.

Kruer, W.L.; Theory and Simulation of Laser Plasma Coupling - I/II. Lectures given at the 20./24. Scottish Universities Summer School in Physics, St. Andrews, Scotland, 1979/1982.

Silin, V.P.; Absorption of Radiation by a Turbulent Laser-Produced Plasma (in Russian). Usp. Fiz. Nauk 145, 225(1985).

Spatschek, K.-H.; Parametric Instabilities in Plasmas (in German). Fortschr. d. Physik 24, 687(1976).

Wong, A.Y.; Electromagnetic Wave Interactions with Inhomogeneous Plasmas. UCLA Report No. PPG 277, 1977.

Part I Basic equations

1 Hydrodynamic plasma description

The analysis in this study is based on the two-fluid model of a plasma. Hence the basic equations are the continuity equation, the equation of motion and an equation of state which controls the dynamic change of the pressure, for the separate ionic and electronic constituents of the plasma:

$$\frac{\partial N_\alpha}{\partial t} + \text{div}\,(N_\alpha \vec{v}_\alpha) = 0, \tag{1.1}$$

$$m_\alpha N_\alpha \frac{\partial \vec{v}_\alpha}{\partial t} + (\vec{v}_\alpha \text{grad})\,\vec{v}_\alpha + \text{grad}\, p_\alpha - q_\alpha (\vec{E} + \vec{v}_\alpha \times \vec{B}) = \vec{F}_\alpha, \tag{1.2}$$

$$\text{grad}\, p_\alpha = \gamma_\alpha T_\alpha\, \text{grad}\, N_\alpha. \tag{1.3}$$

The equations are written for ions and electrons, according as one choses the subscript $\alpha = e$, $\alpha = i$. m_α, N_α, $\vec{v}_\alpha$, p_α, T_α, q_α represent, respectively, the mass, the number density, the velocity, the scalar pressure, the temperature and the charge density of the fluid consisting of particles of sort α. We have $q_e = -eN_e$, $q_i = ZeN_i$ where e is the magnitude of the charge of the electron and Z is the ionic charge number.
The term F describes the momentum exchange between the two fluids, given by the formula

$$\vec{F}_e = -\vec{F}_i = -N_e m_e \nu_{ei} (\vec{v}_e - \vec{v}_i) \tag{1.4}$$

where ν_{ei} denotes the electron-ion collision frequency. For the electron gas there is no general coefficient γ_e valid for arbitrary time scales of the electron density variation. In order to salvage the simple concept of describing the electrons by an equation of state we have to distinguish between rapid and slow changes of the electron density. A process in which the electron density varies on the same time scale as the ion density is referred to as quasi-stationary. Owing to the high thermal conducticity of the electron gas, an isothermal equation of state ($\gamma_e = 1$) is appropriate in this situation. Rapid changes of the electron density are understood as taking place on such a short time scale that the ions can be considered as being immobile. Such time scale is given, for example, by the reciprocal $1/\omega_p$ of the electron plasma frequency $\omega_p = (e^2 N_e/\varepsilon_o m_e)^{1/2}$. In a certain approximation the electrons then behave like a fluid with the coefficient $\gamma_e = 3$ as for one-dimensional processes.

Eqs. (1.1) - (1.3) have to be completed by the Maxwell equations

$$\mathrm{rot}\,\vec{E} = -\frac{\partial B}{\partial t}, \tag{1.5}$$

$$\mathrm{rot}\,\vec{B} = \mu_0 \vec{j} + \frac{1}{c^2}\frac{\partial \vec{E}}{\partial t}, \tag{1.6}$$

$$\mathrm{div}\,\vec{E} = \frac{q}{\varepsilon_0}, \tag{1.7}$$

$$\mathrm{div}\,\vec{B} = 0 \tag{1.8}$$

with the current density

$$\vec{j} = e(ZN_i\vec{v}_i - N_e\vec{v}_e) \tag{1.9}$$

and the charge density

$$q = e(ZN_i - N_e)\,. \tag{1.10}$$

The macroscopic plasma equations presented above can be derived from the Boltzmann equation

$$\frac{\partial f_\alpha}{\partial t} + \vec{u}\,\frac{\partial f_\alpha}{\partial \vec{r}} + \frac{e_\alpha}{m_\alpha}\left[\vec{E} + \vec{u} \times \vec{B}\right]\frac{\partial f_\alpha}{\partial \vec{u}} = \left(\frac{\partial f_\alpha}{\partial t}\right)_{\mathrm{coll.}} \tag{1.11}$$

which describes the change in the particle velocity distribution function $f_\alpha(\vec{r}, \vec{u}, t)$ under the influence of electric and magnetic fields and collisions between the particles. The solution of Eq. (1.11) is rather complicated, even for relatively simple cases. The transition to a continuum description goes along with a considerable mathematical simplification and is thus of great importance. The formulation of general criterions of its validity and applicability, however, is rather difficult.

The formal way to derive macroscopic equations from the Boltzmann equation is well-known (see, for example /1/). They result as equations for the moments

$$N_\alpha = \int f_\alpha d\vec{u}\,,\quad \vec{v}_\alpha = \frac{1}{N_\alpha}\int \vec{u}\, f_\alpha d\vec{u}\,,\quad \vec{\vec{p}}_\alpha = \int (\vec{w}_\alpha \vec{w}_\alpha)\, f_\alpha d\vec{u}$$

($\vec{w}_\alpha = \vec{u} - \vec{v}_\alpha$), whereby the equation for a certain moment involves the moment of the next higher order. Hence, in order to close the system assumptions must be made on the velocity distribution function or on the higher moments.

In a sufficiently collision-dominated system the velocity distribution function is close to a local Maxwellian

$$f_{\alpha o}(\vec{u}) = N_\alpha \left(\frac{m_\alpha}{2\pi k_B T_\alpha}\right)^{3/2} \exp - \left\{\frac{(\vec{u} - \vec{v}_\alpha)^2}{2k_B T_\alpha / m_\alpha}\right\} \tag{1.12}$$

(k_B Boltzmann constant). This has in particular the consequence that the pressure tensor $\overleftrightarrow{p}_\alpha$ is approximately isotropic

$$\overleftrightarrow{p}_\alpha = p_\alpha \overleftrightarrow{I}$$

($\overleftrightarrow{I}$ unit tensor) with

$$p_\alpha = \frac{1}{3} m \int \vec{w}_\alpha^2 f_{\alpha o} d\vec{u} = N_\alpha k_B T_\alpha \tag{1.13}$$

According to an appropriate assumption on the heat flow, the system of continuity equation and momentum equation may be closed by either an adiabatic equation of state

$$\frac{d}{dt}\left(p_\alpha N_\alpha^{-\gamma_\alpha}\right) = 0 \tag{1.14}$$

with, for example $\gamma_\alpha = 5/3$ (no internal degrees of freedom), or an isothermal one,

$$\frac{d}{dt}\left(p_\alpha N_\alpha^{-1}\right) = 0 . \tag{1.15}$$

The approximation of a local Maxwellian distribution function $f_{\alpha o}$ is justified as long as the length l and the time τ characteristic of distances and times over which macroscopic plasma quantities change appreciably, are large compared to the mean free path l_α and the relaxation time τ_α towards $f_{\alpha o}$, respectively; i. e.,

$$\tau \gg \tau_\alpha , \tag{1.16}$$

$$l \gg l_\alpha . \tag{1.17}$$

Thus, these inequalities are an essential requirement for the validity of a hydrodynamic description.

In a fully ionized plasma, however, where collisions are largely electrostatic in nature, τ_α and l_α may become large quantities and conditions (1.16), (1.17) may be difficult to satisfy. This is particularly true for processes in the electron gas taking place on the time scale of the reciprocal $1/\omega_p$ of the electron plasma frequency. In this case, condition (1.16) is violated and the description of the electron gas by a macroscopic model is questionable.

This, however, does not necessarily imply that a continuum description is completely inapplicable to the electrons. Kinetic theory reveals that even in collisionless plasmas electrons may exhibit a fluid-like behaviour due to the generation of self-consistent fields which produce a coherent motion. Rapid changes in the electron gas on the time scale $1/\omega_p$ can be described by a hydrodynamic approach presupposed that the condition

$$l \gg \lambda_D \tag{1.18}$$

is satisfied where

$$\lambda_D = \left(\mathcal{E}_o k_B T_e / e^2 N_e \right)^{1/2} \tag{1.19}$$

is the Debye length /2/. The electron gas then reacts according to an equation of state

$$\frac{d}{dt}\left(p_e N_e^{-3} \right) = 0 \tag{1.20}$$

with a coefficient $\gamma_e = 3$. The question of the accuracy of this equation and the character of the approximation with which it is connected, however, can consistently be assessed only on the basis of a kinetic analysis.

The above system consisting of the hydrodynamic equations for electrons and ions and Maxwell's equations can be specified to a form appropriate for the study of either high-frequency or low-frequency processes. This will be done in the next chapter.

References

/1/ Spitzer, L.; Physics of Fully Ionized Gases. 2. ed., Intersc. Publ., New York-London 1962.

/2/ Jackson, J.D.; Plasma Phys. 1, 171(1960).

2 Basic equations for high-frequency processes

In this chapter we shall adapt the basic equations presented in the preceding section to the case of high-frequency processes. Characteristic feature of these processes is that only the electrons are considered to respond to the external applied field. Ions are treated as to form an immobile charge background. The equations for the ion dynamics can then be omitted and the basic system reduces essentially to a nonlinear wave equation, similar to the basic equation for nonlinear optics. We shall present several steps of its derivation.

We first consider a situation in which a nonuniform ion fluid is in equilibrium with a hot electron gas, i. e., $\vec{v}_e = \vec{v}_i = 0$, $N_i = N_i(\vec{r})$. The associated steady state electron density N_{eo} is then also nonuniform and a static electric field $\vec{E}_o$ is established which compensates the pressure gradient of the electron fluid. Using $p_e = N_{eo} k_B T_e$ from the electron momentum Equation (1.2) it follows

$$k_B T_e \operatorname{grad} N_{eo} + eN_{eo} \vec{E}_o = 0 ,$$

i. e.,

$$\vec{E}_o = - \frac{k_B T_e}{eN_{eo}} \operatorname{grad} N_{eo} . \tag{2.1}$$

On the other hand, $\vec{E}_o$ and N_{eo} are linked via the Poisson Equation (1.7) by

$$\operatorname{div} \vec{E}_o = \frac{e}{\mathcal{E}_o} (ZN_i - N_{eo}) .$$

This, together with Eq. (2.1) allows an estimation, to what extent the static electron density may deviate from the density ZN_i of positive charges:

$$\left| N_{eo} - ZN_i \right| = \frac{\mathcal{E}_o}{e} \left| \operatorname{div} \vec{E}_o \right| = \frac{\mathcal{E}_o k_B T_e}{e^2 N_e} \left| \Delta N_{eo} \right| \propto \frac{\lambda_D^2}{L_o^2} N_{eo} , \tag{2.2}$$

L_o is the scale lenght of the ion density gradient. Thus, the relative density difference between negative and positive charges is small to second order in the parameter λ_D / L_o.
That is, a significant distortion of the quasineutrality of a plasma under steady state conditions is possible only over distances of the order of the Debye length λ_D. We restrict ourselves to

the case $L_o \gg \lambda_D$ and use the approximation of quasineutrality

$$N_{eo} = Z N_i \tag{2.3}$$

Now we study the electron dynamics of such a non-uniform equlibrium configuration when a high-frequency electric field is present. In the ansatz

$$N_e = N_{eo} + n_e \tag{2.4}$$

n_e accounts for rapid changes in the electron density. Eq. (1.2) for the electrons yields with Eq. (2.4)

$$\frac{\partial \vec{v}_e}{\partial t} + (\vec{v}_e \,\mathrm{grad})\vec{v}_e + \nu_{ei}\vec{v}_e + \frac{k_B T_e}{m_e N_e}\left\{3\,\mathrm{grad}\, n_e - \frac{n_e}{N_{eo}}\,\mathrm{grad}\, N_{eo}\right\}$$

$$+ \frac{e}{m_e}(\vec{E} + \vec{v}_e \times \vec{B}) = 0 \tag{2.5}$$

This equation, together with the Maxwell equations

$$\mathrm{rot}\,\vec{E} = -\frac{\partial \vec{B}}{\partial t}, \tag{2.6}$$

$$\mathrm{rot}\,\vec{B} = \frac{1}{c^2}\frac{\partial \vec{E}}{\partial t} - \mu_o e N_e \vec{v}_e, \tag{2.7}$$

$$\mathrm{div}\,\vec{E} = -\frac{e n_e}{\varepsilon_o}, \tag{2.8}$$

$$\mathrm{div}\,\vec{B} = 0 \tag{2.9}$$

and Eq. (2.4) govern the electron dynamics. The continuity equation for the electrons need not be included explicitely since the Maxwell equations already contain it. This is seen by applying the operation div on Eq. (2.7) and rewriting it with the help of the Poisson equation.

In order to reduce Eq. (2.5) to a more convenient form, let us first estimate the order of magnitude of the different terms in the equation of motion of the electrons.
Neglecting nonlinear terms, collisions and assuming cold electrons this equation reduces to the simple form

$$\frac{\partial \vec{v}_e}{\partial t} + \frac{e}{m_e}\vec{E} = 0 \tag{2.10}$$

which is well-known from the single particle dynamics (motion of an electron in the field $\vec{E}$). These two terms are regarded as leading terms in Eq. (2.5) having the same order of magnitude. With Ω and L being, respectively, a characteristic frequency and a characteristic length for the variation of $\vec{v}_e$, $\vec{E}$, we get for the nonlinear terms the estimation

$$\frac{|(\vec{v}_e \operatorname{grad})\vec{v}_e|}{|\partial\vec{v}_e/\partial t|}, \; \frac{e}{m_e}\frac{|\vec{v}_e \times \vec{B}|}{|\partial\vec{v}_e/\partial t|} \quad \propto \quad \frac{v_E}{\Omega \; L}$$

with $v_E = \max|\vec{v}_e|$. The product $\Omega\ L$ may be interpreted, for instance, as phase velocity of a propagating disturbance. Having in mind an electromagnetic disturbance with $v_{ph} \approx c$, the ratio v_E/c appears to be a measure for the influence of nonlinear effects in the electron continuum. For longitudinal perturbations we have $v_{ph} \gtrsim v_T$ and nonlinear effects become important when v_E/v_T approaches one $(v_T = (2k_BT_e/m_e)^{1/2}$ thermal electron velocity).

The terms proportional to T_e which account for the influence of the thermal electron motion on the electron dynamics can be estimated as

$$\frac{k_BT_e|\operatorname{grad} n_e|}{m_eN_e} \Bigg/ \frac{e}{m_e}|\vec{E}| \propto \left(\frac{\lambda_D}{L}\right)^2 .$$

That means, that due to the assumption expressed by Eq. (1.18) one can expect that these terms cause only a modification of the behaviour described by the simplified Eq. (2.10). The term proportional to grad N_{eo} itself is small of the order of λ_D/L compared with the first term in the bracket and thus will be neglected subsequently. This, of course, does not imply the transition to a homogeneous medium, because the equilibrium density $N_{eo}(r)$ still appears in the equation.

The factor $1/N_e$ in Eq. (2.5) is the source of another nonlinear term. Expanding it as

$$\frac{1}{N_e} = \frac{1}{N_{eo} + n_e} \approx \frac{1}{N_{eo}}\left(1 - \frac{n_e}{N_{eo}}\right)$$

assuming $|n_e| \ll N_{eo}$, the fourth term in Eq. (2.5) reads

$$\frac{3k_BT_e}{m_eN_e}\operatorname{grad} n_e \approx \frac{3k_BT_e}{m_eN_{eo}}\operatorname{grad} n_e - \frac{3k_BT_en_e}{m_eN_{eo}^2}\operatorname{grad} n_e .$$

Compared with the other nonlinear terms, the second term in this equation is of the order

$$\frac{\dfrac{3k_BT_e}{m_eN_{eo}^2}|n_e|\,|\operatorname{grad} n_e|}{|(\vec{v}_e \operatorname{grad})\vec{v}_e|} \propto 3\frac{\Omega^2}{\omega_{pe}^2}\left(\frac{\lambda_D}{L}\right)^2$$

and, thus, will be neglected in the further analysis.

Hence, we arrive at the following equation of motion for the electrons

$$\frac{\partial\vec{v}_e}{\partial t} + (\vec{v}_e \operatorname{grad})\vec{v}_e + \nu_{ei}\vec{v}_e + \frac{3k_BT_e}{m_eN_{eo}}\operatorname{grad} n_e + \frac{e}{m_e}(\vec{E} + \vec{v}_e \times \vec{B}) = 0$$

This equation can further be reduced by combining the two remaining nonlinear terms. Using the vector identity $(\vec{a}\,\mathrm{grad})\vec{a} = 1/2\,\mathrm{grad}\,(\vec{a} \times \vec{a}) - \vec{a} \times \mathrm{rot}\,\vec{a}$ we can write

$$(\vec{v}_e \mathrm{grad})\,\vec{v}_e + \frac{e}{m_e}(\vec{v}_e \times \vec{B}) = \frac{1}{2}\,\mathrm{grad}\,(\vec{v}_e \cdot \vec{v}_e) - \vec{v}_e \times \mathrm{rot}\,(\vec{v}_e - \frac{e}{m_e}\vec{B})\,.$$

Substituting this expression into the above equation of motion, taking rot of the resulting equation and using Eq. (2.6) gives

$$\frac{\partial}{\partial t}\left(\mathrm{rot}\,\vec{v}_e - \frac{e}{m_e}\vec{B}\right) = \mathrm{rot}\left\{\vec{v}_e \times \left(\mathrm{rot}\,\vec{v}_e - \frac{e}{m_e}\vec{B}\right)\right\}.$$

That is, the term $\mathrm{rot}\,\vec{v}_e - e/m_e\vec{B}$ can be removed by setting it equal to zero at the initial time.

Eliminating finally n_e through the Poisson equation, the equation for the electron dynamics takes the form

$$\frac{\partial \vec{v}_e}{\partial t} + \nu_{ei}\vec{v}_e - \frac{3k_BT_e}{m_eN_{eo}}\varepsilon_o\,\mathrm{grad\,div}\,\vec{E} + \frac{e}{m_e}\vec{E} + \mathrm{grad}\,(\vec{v}_e \cdot \vec{v}_e) = 0 \tag{2.11}$$

The Maxwell equation (1.5), (1.6) can be combined in the usual way to yield the wave equation

$$\frac{\partial^2\vec{E}}{\partial t^2} + c^2\,\mathrm{rotrot}\,\vec{E} = -\mu_o\frac{\partial \vec{j}}{\partial t} = \mu_o\frac{\partial}{\partial t}(N_e\vec{v}_e) = \mu_o N_{eo}\frac{\partial \vec{v}_e}{\partial t} + \frac{\partial}{\partial t}(n_e\vec{v}_e)\,.$$

In nonlinear optics, the right-hand side is usually expressed in terms of the generalized polarization $\vec{P} = \vec{P}(\vec{E})$ via the relation $\vec{j} = \partial\vec{P}/\partial t$ containing both linear and nonlinear contributions. Here we substitute $\partial\vec{v}_e/\partial t$ with the help of the electron momentum equation and rewrite the wave equation by collecting linear terms on the left-hand side leaving only the nonlinear terms on the right-hand side.

This gives

$$-\frac{\partial^2\vec{E}}{\partial t^2} - c^2\,\mathrm{rotrot}\,\vec{E} + 3v_T^2\,\mathrm{grad\,div}\,\vec{E} - \omega_p^2\vec{E} - \frac{m_e}{e}\nu_{ei}\,\omega_p^2\vec{v}_e = \vec{Q}_{NL} \tag{2.12}$$

with

$$\vec{Q}_{NL} = \frac{m_e}{e}\,\omega_p^2\,\frac{1}{2}\,\mathrm{grad}\,(\vec{v}_e \cdot \vec{v}_e) + \frac{\partial}{\partial t}(\vec{v}_e\,\mathrm{div}\,\vec{E}) \tag{2.13}$$

This equation is the goal of this section. It governs the interaction of a high-frequency electric field with a nonuniform warm electron continuum. The source term $\vec{Q}_{NL}$ on the right-hand side gives rise to a nonlinear response of the electron gas to an external applied electromagnetic wave. The term $\mathrm{grad}\,(\vec{v}_e \cdot \vec{v}_e)$ results, as has been shown, from the equation of motion of the electrons while the second nonlinearity originates from the product $n_e\vec{v}_e$ in the expression for the current density. The term $3v_T^2$ graddiv $\vec{E}$ represents the influence of the finite electron temperature; the plasma inhomogeneity enters the theory via terms containing $\omega_p^2 \propto N_{eo}$.

The nonlinear terms (and the collision term), however, still contain the electron velocity $\vec{v}_e$. It should be eliminated, in principle, again by the electron momentum equation (2.11).

We may, however, utilize a simplified relation between $\vec{v}_e$ and $\vec{E}$ for that purpose. In particular, we may take the reduced equation of motion

$$\frac{\partial \vec{v}_e}{\partial t} = -\frac{e}{m_e}\vec{E} \qquad (2.14)$$

to carry out the elimination. This would imply, generally speaking, the restriction to nonlinearities of second order in the electric field.

In a higher degree of accuracy, however, the relation between $\vec{v}_e$ and $\vec{E}$ itself is nonlinear. When, for example, Eq. (2.14) is improved by including the gradient term, e. g., when using the version

$$\frac{\partial \vec{v}_e}{\partial t} = -\frac{e}{m_e}\vec{E} - \frac{1}{2}\,\mathrm{grad}\,(\vec{v}_e \cdot \vec{v}_e) \qquad (2.15)$$

of the equation of motion for the electrons to eliminate $\vec{v}_e$ from Eq. (2.12), this procedure can be accomplished only by repeatedly going back to Eq. (2.15) until the cycle is closed by using Eq. (2.14) as final step. From this consideration we can conclude that the source term $\vec{Q}_{NL}$ in principle expresses nonlinearities of arbitrary high order in the electric field strength $\vec{E}$.

3 Basic equations for low-frequency processes

In the previous Chapter we derived a nonlinear wave equation capable of describing dynamic processes of the electron fluid under the action of a high-frequency electromagnetic wave. Now we apply the basic system to processes in which the ion dynamics is involved.

Thereby, we concentrate on a specific nonlinear process, namely, the nonlinear modification of the density profile produced by the ponderomotive force associated with the high-frequency electric fields in the plasmas. The ponderomotive force is essentially proportional to the gradient of the electromagnetic energy density and represents the radiation pressure effects that can be produced by an intense electromagnetic wave.

Hora /1/ pointed first to the relevance of this force in the context of laser-plasma interaction. The aim of this Chapter is to derive equations for the plasma dynamics including the ponderomotive force of the high-frequency field.

Introducing the mass velocity

$$\vec{v} = \frac{m_i \vec{v}_i N_i + m_e \vec{v}_e N_e}{m_e N_e + m_i N_i} \tag{3.1}$$

and the current density

$$\vec{j} = e(Z N_i \vec{v}_i - N_e \vec{v}_e) \tag{3.2}$$

instead of the velocities $\vec{v}_i$, $\vec{v}_e$, two rather complicated equations for $\vec{v}$, $\vec{j}$ can be derived from from the fluid equations /2/. One of them has the character of a equation of motion of the plasma while the other represents the so called generalized Ohm's law. In the presence of a high-frequency electromagnetic field these equations contain slowly varying as well as rapidly varying quantities. In order to arrive at equations for the plasma as a whole, we have to average them over the time scale associated with the electrons. This procedure was performed by Gorbunov /3/ with the essential result that the terms quadratic in the rapidly varying quantities survive the averaging process and produce residues, one of which describes the ponderomotive force of a high-frequency field on the plasma.

In our derivation we assume the presence of a monochromatic high-frequency field (frequency ω) and represent the rapidly varying quantities $\vec{E}$, $\vec{v}_e$, n_e in the form

$$\vec{E}(\vec{r}, t) = \vec{E}_o(\vec{r}, t) + \mathrm{Re}\left\{\vec{E}_\omega(\vec{r}, t)\, e^{-i\omega t}\right\},$$

$$\vec{v}_e(\vec{r}, t) = \mathrm{Re}\left\{\vec{v}_{e\omega}(\vec{r}, t)\, e^{-i\omega t}\right\},$$

$$n_e(r, t) = \mathrm{Re}\left\{n_{e\omega}(\vec{r}, t)\, e^{-i\omega t}\right\}. \tag{3.3}$$

Thereby, $\vec{E}_\omega$, $\vec{v}_{e\omega}$, $n_{e\omega}$ are complex quantities that vary slowly on the time scale $1/\omega$. $\vec{E}_o$ refers to a quasistatic field similar to that given by Eq. (2.1). The dc electron velocity $\vec{v}_{eo}$ has been ignored in Eq. (3.3). This implies that neither a dc current nor an electron beam is assumed to be present in the background plasma.

The ion equations do not contain rapidly varying quantities because we assume that the ions are unable to respond to the high frequency field owing to their high mass. Thus, the equation of motion for the ions reduces to

$$\frac{\partial \vec{v}_i}{\partial t} + (\vec{v}_i \,\mathrm{grad})\, \vec{v}_i + \frac{1}{m_i N_i}\,\mathrm{grad}\, p_i - Z\frac{e}{m_i}\vec{E}_o = 0 \tag{3.4}$$

(no dc magnetic field).

The contribution of the electrons to the plasma dynamics can be derived from Eq. (2.11). Introducing the ansatz (3.3) into this equation and treating terms $\propto \exp(-i\omega t)$ and dc-terms separately yields the two equations

$$\frac{\partial \vec{v}_e}{\partial t} + \nu_{ei}\vec{v}_e - \frac{3v_T^2}{eN_{eo}}\,\mathcal{E}_o\,\mathrm{grad}\,\mathrm{div}\,\vec{E} + \frac{e}{m_e}\vec{E} = 0 \tag{3.5}$$

and

$$\frac{v_T^2}{N_{eo}}\,\mathrm{grad}\, N_{eo} + \frac{e}{m_e}\vec{E}_o + \frac{1}{4}\,\mathrm{grad}\,(\vec{v}_{e\omega}\cdot\vec{v}^{\,*}_{e\omega}) = 0. \tag{3.6}$$

An asterix superscript on a complex quantity denotes the complex conjugate.
The last equation for the zero-frequency terms, rewritten as

$$\frac{e}{m_e}\vec{E}_o = -\frac{k_B T_e}{m_e N_{eo}}\,\mathrm{grad}\, N_{eo} - \frac{1}{4}\,\mathrm{grad}\,|\vec{v}_{e\omega}|^2 \tag{3.7}$$

may be understood as a modification of Eq. (2.1). The additional term $\propto \mathrm{grad}|\vec{v}_{e\omega}|^2$ results from the self-interaction of the rapidly varying electric field (combination of termes $\propto \exp(-i\omega t)$ with terms $\propto \exp(+i\omega t)$) and represents a dc-force caused by this field. Introducing Eq. (3.7) into the momentum balance for the ions gives

$$\frac{\partial \vec{v}_i}{\partial t} + (\vec{v}_i\,\mathrm{grad})\,\vec{v}_i = -\frac{\gamma_i k_B T_e}{m_i N_i}\,\mathrm{grad}\, N_i - \frac{Z k_B T_e}{m_e N_{eo}}\,\mathrm{grad}\, N_{eo} - \frac{1}{4} Z\frac{m_e}{m_i}\,\mathrm{grad}\,|\vec{v}_{e\omega}|^2 .$$

This equation can be simplified as follows: First, we make use of the quasineutrality approximation (Eq. (2.3)) and set $N_{eo} = ZN_i = N$. Further, we identify $\vec{v}_i$ with the mass velocity $\vec{v}$ of the plasma. As may be seen from Eq. (3.1), for a quasineutral plasma this is accurate to small terms of the order of m_e/m_i, presupposed that $\vec{v}_{eo}$ is of the same order as $\vec{v}_i$. Finally, we eliminate $\vec{v}_e$ in favour of $\vec{E}_\omega$ by using the simplified equation of motion in the form

$$i\,\omega \vec{v}_e = \frac{e}{m_e}\vec{E}_\omega . \tag{3.8}$$

This combines to yield the following equation of motion for the plasma

$$\frac{\partial \vec{v}}{\partial t} + (\vec{v}\,\mathrm{grad})\,\vec{v} = -\frac{Zk_B T_e + \gamma_i k_B T_i}{m_i}\,\frac{1}{N}\,\mathrm{grad}\,N - \frac{1}{4}\,\frac{Ze^2}{\omega^2 m_e m_i}\,\mathrm{grad}\,\left|\vec{E}_\omega\right|^2 . \tag{3.9}$$

The first term on the right-hand side represents the force on the plasma that is caused by a density gradient. This term keeps to be significant even when the ion temperature T_i is much less than the electron temperature T_e. The second term represents the ponderomotive force of the high frequency field, caused by a gradient of its energy density. This term was first derived in /4, 5/. Subsequently derivations were also given in /6 –8/.

Eq. (3.9) has to be completed by an equation for the complex amplitude $\vec{E}_\omega$. Representing the high-frequency field as

$$\vec{E}(\vec{r}, t) = \mathrm{Re}\left\{\vec{E}_\omega(\vec{r}, t)\, e^{-i\omega t}\right\} \tag{3.10}$$

and assuming

$$\frac{\partial^2 \vec{E}_\omega}{\partial t^2} \ll \omega^2 \left|\vec{E}_\omega(\vec{r}, t)\right| ,$$

we get

$$\frac{\partial^2}{\partial t^2}\vec{E}(\vec{r}, t) \cong \mathrm{Re}\left[\left(-2i\omega\frac{\partial \vec{E}_\omega}{\partial t} - \omega^2 \vec{E}_\omega\right) e^{-i\omega t}\right] .$$

Introducing this into Maxwell's equations and using in addition the denotation

$$\mathcal{E}(\omega, N) = 1 - \frac{\omega_p^2}{\omega^2}\left(1 - i\frac{\nu_{ei}}{\omega}\right) = 1 - \frac{N}{N_c}\left(1 - \frac{\nu_{ei}}{\omega}\right) \tag{3.11}$$

yields the desired equation

$$2i\omega\frac{\partial \vec{E}_\omega}{\partial t} - c^2\,\mathrm{rot\,rot}\,\vec{E}_\omega + 3v_T^2\,\mathrm{graddiv}\,\vec{E}_\omega + \omega^2\,\mathcal{E}(\omega, N)\,\vec{E}_\omega = 0 \tag{3.12}$$

This equation is frequently referred to as wave equation in modulation representation. N_c denotes the critical density associated with the frequency ω, defined by $\omega = \omega_p$ and given by

$$N_c = \frac{\mathcal{E}_o m_e \omega^2}{e^2} . \tag{3.13}$$

Finally, to close the system we need the equation of continuity for the ions

$$\frac{\partial N_i}{\partial t} + \mathrm{div}\,(N_i \vec{v}_i) = 0$$

which is identical with the mass balance for the plasma owing to $N_i = N$, $\vec{v} = \vec{v}_i$.

The complete system of equations that describes the plasma dynamics under the influence of a high-frequency field now reads

$$\frac{\partial N}{\partial t} + \mathrm{div}\,(N\vec{v}) = 0, \tag{3.14}$$

$$\frac{\partial \vec{v}}{\partial t} + (\vec{v}\,\mathrm{grad})\,\vec{v} = -c_s^2 \frac{1}{N}\,\mathrm{grad}\,N - \frac{1}{4}\frac{Ze^2}{m_e m_i \omega^2}\,\mathrm{grad}\,|\vec{E}_\omega|^2, \tag{3.15}$$

$$2i\omega\frac{\partial \vec{E}_\omega}{\partial t} - c^2\,\mathrm{rotrot}\,\vec{E}_\omega + 3v_T^2\,\mathrm{graddiv}\,\vec{E}_\omega + \omega^2\,\varepsilon(\omega, N)\,\vec{E}_\omega = 0. \tag{3.16}$$

Here,

$$c_s = \left\{\frac{k_B(\gamma_i T_i + ZT_e)}{m_i}\right\}^{1/2} \tag{3.17}$$

is the ion acoustic velocity.

The system (3.14) – (3.16) can be found in similar form in /3/. It describes a coupling between hydrodynamic and electrodynamic processes in the plasma. This coupling is provided first by the ponderomotive force in the momentum balance of the plasma, which gives rise to spatial and temporal variations in density and velocity. The second coupling term is the term $\varepsilon(\omega, N)\,\vec{E}_\omega$ in the modulation representation of the wave equation. It expresses the reaction of density changes back to the spatial field structure via the density dependent dielectrics $\varepsilon(\omega, N)$ of the plasma.

In deriving the equation of motion of the plasma as a whole, the crucial term proved to be the term $\propto \mathrm{grad}\,(\vec{v}_e \cdot \vec{v}_e)$ in the equation of motion for the electrons. Zero-frequency actions originating from nonlinear terms in the expression for the current density or the continuity equation for the electrons need not be taken into consideration. They play a role, e. g., in the context of dc magnetic field generation.

References

/1/ Hora, H., Pfirsch, D., Schlüter, A.; Z. Naturf. 22A, 278(1967).

/2/ Lüst, R.; Fortschr. d. Physik 7, 503(1959).

/3/ Gorbunov, L.M.; Usp. Fiz. Nauk 109, 631(1973).

/4/ Gapunov, A.V., Miller, M.A.; Zh. Eksp. Teor. Fiz. 34, 242(1958).

/5/ Pitaevski, L.P.; Zh. Eksp. Teor. Fiz. 39, 1450(1960).

/6/ Hora, H.; Phys. Fluids 12, 182(1969).

/7/ Lindl, J.O., Kaw, P.K.; Phys. Fluids 14, 371(1971).

/8/ Shearer, J.W., Eddleman, J.L.; Phys. Fluids 16, 1753(1973).

Part II Elements of linear wave propagation

4 Linear wave propagation in plasmas

Before proceeding to nonlinear phenomena we outline briefly several features of linear wave propagation in a nonuniform plasma. The amplitude of a wave in an inhomogeneous plasma may undergo large spatial variations due to zeros in the index of refraction which cause reflection and a superposition of oppositely propagating waves, or due to linear mode conversion. The wave form must be calculated carefully before the amplitude can be inserted into any nonlinear theory. Spatial amplitude gradients, for example, are responsible for the action of ponderomotive forces.

Linear wave propagation has been treated extensively and the results are summarized in a number of monographs (see, for example /1 - 3/). We confine ourselves here to a brief description of the main processes as linear mode conversion and resonance absorption.

4.1 Linear wave equation

We consider a monochromatic wave that propagates through an inhomogeneous static plasma with time-independent amplitude. The wave field is represented in the form

$$\vec{E}(\vec{r}, t) = \mathrm{Re}\left\{\vec{E}_\omega(\vec{r})\, e^{-i\omega t}\right\} \tag{4.1}$$

with the complex amplitude $\vec{E}_\omega(\vec{r})$. Introducing this ansatz into the wave equation (2.11) and ignoring nonlinear terms yields the following steady-state wave equation

$$-c^2 \,\mathrm{rotrot}\, \vec{E}_\omega + 3v_T^2 \,\mathrm{graddiv}\, \vec{E}_\omega + \omega^2 \mathcal{E}(r)\, \vec{E}_\omega = 0 \tag{4.2}$$

where the dielectric constant $\mathcal{E}(\vec{r})$ is given by

$$\mathcal{E}(\vec{r}) = 1 - \frac{\omega_p^2(\vec{r})}{\omega^2}\left(1 - i\frac{\nu_{ei}}{\omega}\right) = 1 - \frac{N(\vec{r})}{N_c}\left(1 - i\,\frac{\nu_{ei}}{\omega}\right). \tag{4.3}$$

It is useful to split Eq. (4.2) into two equations by applying the operations rot and div on it. This yields the pair of equations

$$c^2 \triangle \,\mathrm{rot}\, \vec{E}_\omega + \omega^2\, \mathcal{E}(\vec{r})\, \mathrm{rot}\, \vec{E}_\omega - \mathrm{grad}\, \mathcal{E} \times \vec{E}_\omega = 0, \tag{4.4}$$

$$3v_T^2 \triangle \mathrm{div}\, \vec{E}_\omega + \omega^2 \mathcal{E}\, \mathrm{div}\, \vec{E}_\omega - \mathrm{grad}\, \mathcal{E} \cdot \vec{E}_\omega = 0. \tag{4.5}$$

From these equations it is seen that transverse waves (rot $\vec{E}_\omega \neq 0$) and longitudinal waves (div $\vec{E}_\omega \neq 0$) are decoupled in a homogeneous plasma, where grad $\mathcal{E} \propto$ grad N = 0. In this case, we get with $\vec{E}(\vec{r}) \propto \exp(i\vec{k}\vec{r})$ the dispersion relations

$$c^2k^2 = \omega^2 - \omega_p^2\left(1 - i\frac{\nu_{ei}}{\omega}\right) \tag{4.6}$$

for transverse waves and

$$3v_T^2k^2 = \omega^2 - \omega_p^2\left(1 - i\frac{\nu_{ei}}{\omega}\right) \tag{4.7}$$

(Bohm-Gross relation) for longitudinal waves in a plasma. Common to both wave types is that real wavenumbers, i. e., wave transmission, exist only for $\omega > \omega_p$. For $\omega < \omega_p$ k is imaginary and the wave is evanescent. Reflection of the waves take place at the critical density N_c. The phase velocities and group velocities of both wave types are given by

$$V_{ph} = \frac{c}{\left(1 - \omega_p^2/\omega^2\right)^{1/2}}, \quad V_{gr} = c\left(1 - \omega_p^2/\omega^2\right)^{1/2} \tag{4.8}$$

for electromagnetic waves and by

$$V_{ph} = \frac{\sqrt{3}\, v_T}{\left(1 - \omega_p^2/\omega^2\right)^{1/2}}, \quad V_{gr} = \sqrt{3}\, v_T\left(1 - \omega_p^2/\omega^2\right)^{1/2} \tag{4.9}$$

for the longitudinal waves (referred to also as Langmuir waves or plasma waves).

The collision-induced temporal damping decrement for both wave types can be derived from the dispersion relations to be

$$\sigma = -\mathrm{Im}(\omega) = \frac{\omega_p^2}{\omega^2}\,\frac{\nu_{ei}}{2}. \tag{4.10}$$

For longitudinal waves this simplifies to

$$\sigma \approx \frac{\nu_{ei}}{2} \tag{4.11}$$

because $\omega \approx \omega_p$ in this case.

The spatial damping predicted by the dispersion relations is described by the ratio

$$\frac{\mathrm{Im}\, k}{\mathrm{Re}\, k} = \frac{\omega_p^2}{\omega^2 - \omega_p^2} \cdot \frac{1}{2}\,\frac{\nu_{ei}}{\omega}. \tag{4.12}$$

The wavelengths λ_T, λ_L of transverse and longitudinal waves are connected to the vacuum wave length λ_o (all of the same frequency) by the relations

$$\lambda_T = \frac{\lambda_o}{\sqrt{\mathcal{E}}}, \quad \lambda_L = \sqrt{3}\,\frac{v_T}{c}\,\lambda_T = \sqrt{3}\,\frac{v_T}{c}\,\frac{\lambda_o}{\sqrt{\mathcal{E}}}. \tag{4.13}$$

Thus, the wavelength of a longitudinal wave is smaller by the factor $\sqrt{3}\,v_T/c$ than that of the transverse wave of the same frequency.

Kinetic theory predicts a strong increase of the longitudinal wave damping with increasing frequency (Landau damping). The collision term in the dispersion equation (4.7) does not account for this behaviour. This is illustrated in Fig. 4.1, where the ratio Im (k)/Re (k) and the normalized phase velocity v_{ph}/v_T for longitudinal waves are plotted as function of the ratio ω/ω_p.

Graph (1) represents the dependence predicted by Eq. (4.12) for $\nu_{ei} = 0.01$. This graph reveals that usual collisional damping is very effective near the plasma frequency while only little damping is produced for higher frequencies. Graph (2) represents the solution of the kinetic dispersion relation

$$1 - \frac{\omega_p^2}{k^2 v_T^2}\, Z'\left(\frac{\omega}{kv_T}\right) = 0 \tag{4.14}$$

for longitudinal waves in a collisionless plasma with a Maxwellian velocity distribution (Z: plasma dispersion function). In contrast to the prediction of the fluid model, typical feature is here the increase of the damping with growing frequency, which confines the frequency region for propagation to the vicinity of the plasma frequency. Graph (3) presents the result of a simple approximation of the kinetic solution, provided by the introduction of an additional frequency-dependent damping rate ν_L, given by

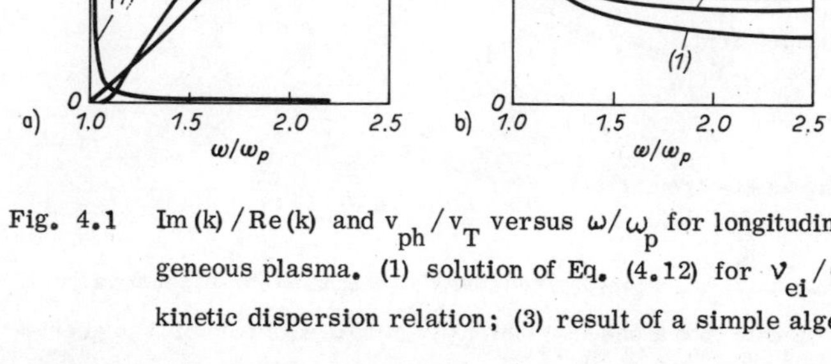

Fig. 4.1 Im (k) / Re (k) and v_{ph}/v_T versus ω/ω_p for longitudinal plasma waves in a homogeneous plasma. (1) solution of Eq. (4.12) for $\nu_{ei}/\omega = 0.01$; (2) solution of the kinetic dispersion relation; (3) result of a simple algebraic approximation

$$\frac{\nu_L}{\omega} = \frac{1}{4}\left(\frac{\omega^2}{\omega_p^2} - 1\right) = \frac{3}{4}\,\frac{k^2\lambda_D^2}{\omega_p^2} \tag{4.15}$$

and by replacing v_T^2 in Eq. (4.7) through

$$v_T^2 \longrightarrow \frac{v_T^2}{1 + i\,\frac{\nu_L}{\omega}} \; . \tag{4.16}$$

We introduce this phenomenological ansatz into the stationary wave equation (4.2) in order to model Landau damping. The damping rate ν_L then damps preferentially the electrostatic part of the high-frequency wave while the usual collision frequency ν_{ei} acts both on transverse and plasma waves. A similar selective damping rate has been introduced in / 4 /. The modified temporal damping decrement for longitudinal waves caused by this phenomenological ansatz is

$$\sigma_L = \frac{\nu_{ei}}{2} + \frac{3}{2}k^2\lambda_D^2\,\nu_L = \frac{1}{2}(\nu_{ei} + \nu_L^*) \tag{4.17}$$

where

$$\nu_L^* = \frac{9}{4}(k\lambda_D)^4 \; . \tag{4.18}$$

In the presence of a density gradient longitudinal and transverse waves are coupled in general. In the case of a weak gradient, when both wave types are of relative independence, Eqs. (4.4), (4.5) can be interpreted as follows: the coupling terms containing the density gradient can be put as small source terms on the right-hand sides.

Eq. (4.4) then describes the excitation of transverse waves due to the simultaneous existence of longitudinal waves and a density gradient. This process is important, e. g., for the interpretation of non-thermal radiation generation. Oppositely, the generation of longitudinal waves by transverse ones according to Eq. (4.5) establishes an "absorption" mechanism that operates independently of collisions. It works most effective near the critical density (see Section 4.3 on resonance absorption).

There are, however, particular situations where no coupling occurs between longitudinal and transverse waves in a nonuniform plasma. A transverse wave that is polarized such that $\mathrm{grad}\, N \cdot \vec{E} = 0$ does not generate longitudinal perturbations. This case is especially realized for propagation along the density gradient.

4.2 Penetration of a plasma by an electromagnetic wave

In this section we apply the linear wave equation derived above to the problem of penetration of an electromagnetic wave from vacuum into a one-dimensionally stratified medium. The geometry of the problem is determined by the constellation of the three vectors (a) the density gradient grad N, (b) the wave vector $\vec{k}$ and (c) the field vector $\vec{E}$ of the wave. We choose a refer-

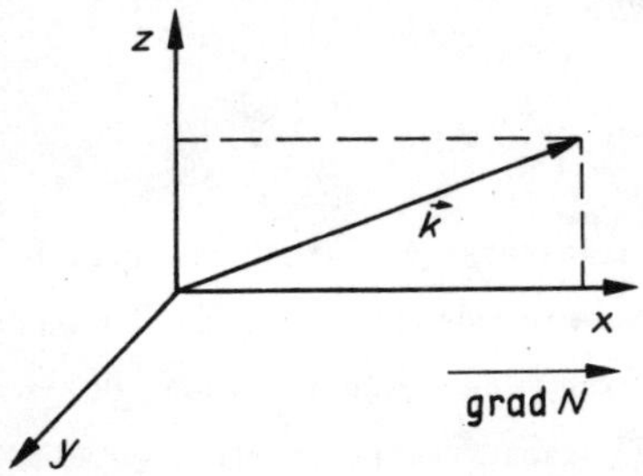

Fig. 4.2 Coordinate system associated with the incidence of an electromagnetic wave (wave vector $\vec{k}$) upon a one-dimensionally stratified plasma

ence system as shown in Fig. 4.2 with grad N in x-direction and the x-z-plane as plane of incidence. The free-space field is assumed to be a plane wave

$$\vec{E}_\omega(x, z) = \vec{E}_o e^{i(k_x x + k_z z)} \tag{4.19}$$

and within the plasma we can write

$$\vec{E}_\omega(x, z) = \vec{E}_\omega(x) e^{ik_z z} . \tag{4.20}$$

The component k_z does not vary with x as a consequence of Snell's law. We set

$$\vec{E}_\omega(x) = (E_x(x), E_y(x), E_z(x)) \tag{4.21}$$

and suppress the index ω on the field components E_x, E_y, E_z. Introducing the expressions (4.20), (4.21) into the wave equation (4.2) we get the system

$$3v_T^2 \frac{d^2E_x}{dx^2} + (3v_T^2 - c^2)\, ik_o \sin\theta \frac{dE_z}{dx} + \omega^2(\varepsilon - \sin^2\theta)E_x = 0 , \tag{4.22}$$

$$c^2 \frac{d^2E_y}{dx^2} + \omega^2(\varepsilon - \sin^2\theta)E_y = 0 , \tag{4.23}$$

$$c^2 \frac{d^2E_z}{dx^2} + ik_o \sin\theta\,(3v_T^2 - c^2)\frac{dE_x}{dx} + \omega^2\left(\varepsilon - 3\frac{v_T^2}{c^2}\right)E_z = 0 \tag{4.24}$$

where $k_o = \omega/c$ is the vacuum wave number and θ is the angle of incidence relative to the density gradient.

This implies the relation

$$k_z = k_o \sin\theta .$$

Obviously, E_y is decoupled from E_x, E_z which indicates two seperated modes. We therefore distinguish between two types of polarization,

(A) s-polarization,
with field vector $\vec{E}$ perpendicular to the plane of incidence ($E_y \neq 0$, $E_x = E_z = 0$);

(B) p-polarization,

with field vector $\vec{E}$ in the plane of incidence (E_x, $E_z \neq 0$, $E_y = 0$).

The s-polarized wave, described by Eq. (4.23) is a purely transverse wave electromagnetic in nature; the corresponding wave equation does not contain terms proportional to v_T^2 and the mode is not affected by finite electron temperature effects. In contrast, the p-polarized wave, described by the pair of Eqs. (4.22), (4.24), is neither purely transverse nor purely longitudinal.

Normal incidence, $\theta = 0$

In this case all three field components can be treated separately. The transverse mode satisfies the wave equation

$$\frac{d^2 E_{y,z}}{dx^2} + k_o^2 \, \mathcal{E}(\omega, N) \, E_{y,z} = 0 \tag{4.25}$$

while the longitudinal mode obeys the equation

$$\frac{d^2 E_x}{dx^2} + \frac{3v_T^2}{c^2} k_o^2 \, \mathcal{E}(\omega, N) \, E_x = 0 . \tag{4.26}$$

Fig. 4.3 presents solutions of Eq. (4.25) for a linear density profile in absence of collisions. As a measure for the steepeness of the density gradient we use the distance L over which the density increases from zero to the critical density N_c, relative to the free space wave length λ_o. The absolute value of the complex field amplitude $|E| = |E_{y,z}|$ is plotted, normalized by the vacuum amplitude E_o. The wave penetrates the plasma until it will be totally reflected near the region of the critical density. Beyond this point the radiation penetrates the plasma in the form of an evanescent electric field. The peaks in $|E|$ result from constructive interference of the incident and the reflected wave.

A typical feature is the amplification of the field with increasing distance of penetration as a consequence of the decreasing group velocity as the wave propagates inwards. It reaches a

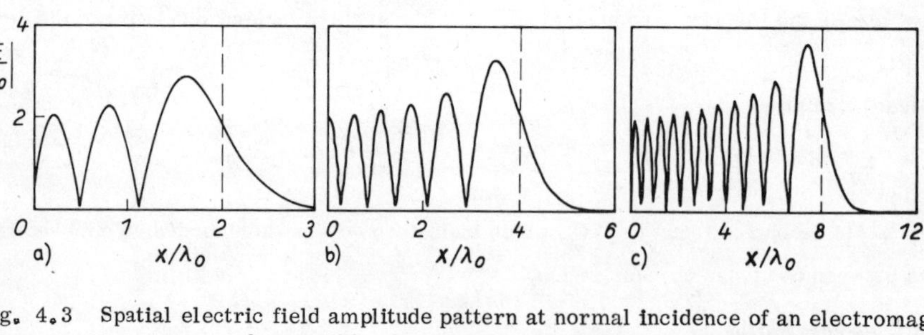

Fig. 4.3 Spatial electric field amplitude pattern at normal incidence of an electromagnetic wave upon a plasma with a linear density gradient of different scale lengths. (a) $L = 2\,\lambda_o$; (b) $L = 4\,\lambda_o$; (c) $L = 8\,\lambda_o$

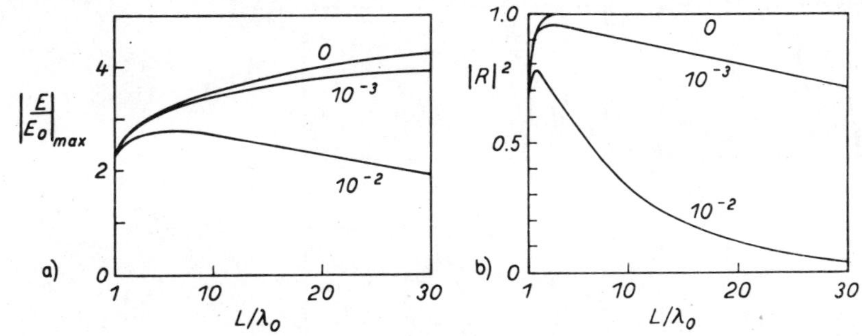

Fig. 4.4 Maximum field amplification (a) and reflection coefficient (b) versus density gradient scale length at normal incidence for different collision frequencies (curve parameter: ν_{ei}/ω)

maximum $|E|_{max}$ in front of the cut-off layer. The amplification factor $|E|_{max}/E_o$ increases slowly with increasing density gradient scale length L. This is seen in Fig. 4.4 (a); the top curve shows $|E|_{max}/E_o$ as function of L/λ_o in absence of collisions.

The spatial field profiles shown in Fig. 4.3 can be expresses in terms of Airy functions. Writing the linear density profile in the form

$$\frac{N(x)}{N_c} = \frac{x}{L} + 1 \quad (-L \leq x) \tag{4.27}$$

the dielectric constant ε is

$$\varepsilon(x) = 1 - \frac{\omega_p^2(x)}{\omega^2} = -\frac{x}{L}.$$

With the help of the substitution

$$\xi = -(k_oL)^{2/3}\frac{x}{L}$$

Eq. (4.25) can be rewritten in the simple form

$$\frac{d^2E}{d\xi^2} + \xi E = 0. \tag{4.28}$$

This equation has the solution

$$E(\xi) = C\,Ai(-\xi) = C\,Ai\left\{(k_oL)^{2/3}\frac{x}{L}\right\} \tag{4.29}$$

where C is a constant determined by fitting to the incoming wave. Concerning details of the analytical theory we refer to /1 - 3/.

The presence of collisions leads to absorption and to a reduction of the field amplification. This is illustrated in Fig. 4.4, which shows the amplification factor $|E|_{max}/E_o$ (a) as well as the reflexion coefficient (b) as function of L/λ_o for $\nu_{ei}/\omega = 0, 10^{-3}, 10^{-2}$.

The longitudinal field described by Eq. (4.26) cannot be excited from outside because there is no coupling to a vacuum field.

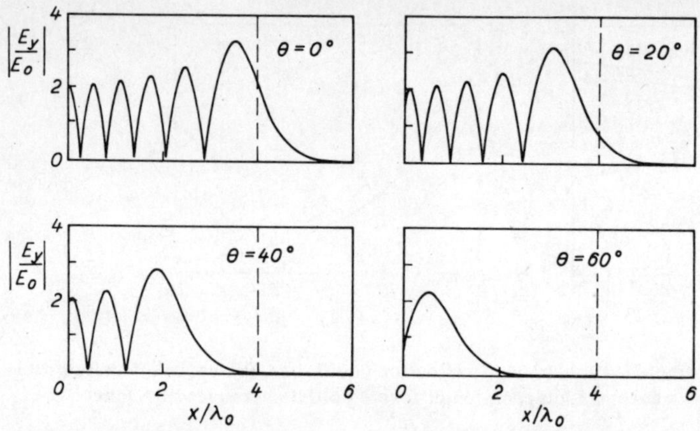

Fig. 4.5 Spatial electric field amplitude pattern at oblique incidence of a s-polarized electromagnetic wave upon a linear density profile ($L = 4\lambda_o$) for different angles of incidence

Oblique incidence, s-polarization

In the case of oblique incidence at an angle θ relative to the density gradient the s-polarized wave obeys the wave equation

$$\frac{d^2E_y}{dx^2} + k_o^2 \left\{ \mathcal{E}(\omega, N) - \sin^2\theta \right\} E_y = 0 \tag{4.30}$$

Because of refraction in the outer layers of the plasma the cut-off now occurs at a lower density N_R than for perpendicular incidence, which is given by

$$\mathcal{E}(\omega, N_R) - \sin^2\theta = 0 .$$

Without collisions this gives

$$N_R = N_c \cos^2\theta .$$

Fig. 4.5 illustrates this situation, showing spatial field profiled for increasing angle of incidence. The wave is seen to be reflected earlier, the standing wave structure is stretched and the field amplification is reduced.

4.3 Resonance absorption

When a p-polarized wave is obliquely incident upon the plasma, the wave amplitude exhibits a fundamentally different behaviour as in the case of s-polarization.

This was pointed out first by Denisov /5/ and Ginzburg /2/. The reason for this is the existence of a field component along the density gradient as long as θ > 0. This component causes an electrostatic charge separation and thus the wave cannot longer remain purely electromagnetic. Similar to the situation for s-polarization, the wave is turned around at the point where $N = N_c \cos^2 \theta$ as predicted by geometric optics. From this layer, the evanescent wave can tunnel through to the critical layer where the nonvanishing longitudinal component of the incident field can resonantly excite plasma waves.

The conversion of an electromagnetic wave into a Langmuir wave establishes an absorption that is not necessarily connected to collisions. It is referred to as resonance absorption.

Resonance absorption has been the subject of various theoretical and experimental studies in the past. Linear theory /2, 5 - 14/ is well understood, most of the work is dealing with nonlinear aspects /15 - 28/ caused by high amplitudes of the excited Langmuir waves (wave breaking, hot electron generation). Experiments have been carried out using microwaves /29 - 31/ as well as laser light /32 - 37/. The presentation here follows mainly the analysis developed in /4/ where resonance absorption was studied on the basis of a warm electron fluid model.

As shown in Section 4.2 the spatial amplitude pattern of E_x, E_z is governed by the pair of equations

$$3v_T^2 \frac{d^2E_x}{dx^2} - ik_o c^2 \sin\theta \frac{dE_z}{dx} + \omega^2(\mathcal{E} - \sin^2\theta)\, E_x = 0, \tag{4.31}$$

$$c^2 \frac{d^2E_z}{dx^2} - ik_o c^2 \sin\theta \frac{dE_x}{dx} + \omega^2\left(\mathcal{E} - \frac{3v_T^2}{c^2}\right) E_z = 0. \tag{4.32}$$

Fig. 4.6 presents a solution of these equations that illustrates the situation when a p-polarized wave is incident upon a linear density profile ($L/\lambda_o = 4$, $v_T/c = 0.05$) with an angle of $\theta = 20^o$. A characteristic feature is the presence of an additional spatial modulation of the

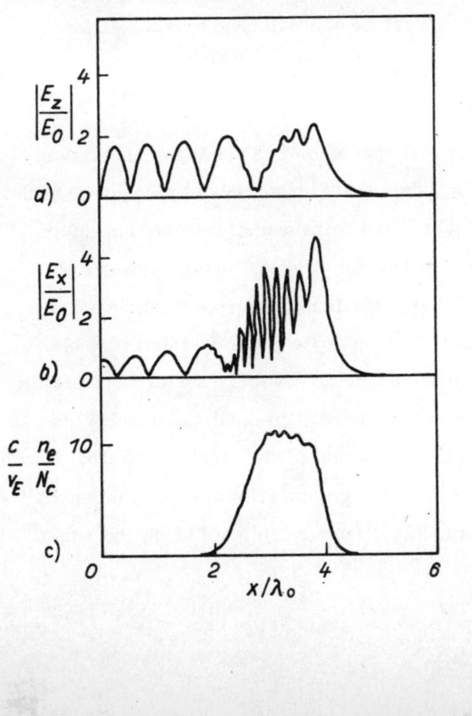

Fig. 4.6 Spatial dependence of the absolute values of the electric field components E_z perpendicular to the density gradient (a) and E_x along the density gradient (b) and the normalized space charge density (c) at incidence of a p-polarized wave upon a linear density profile ($L = 4\lambda_o$, $\theta = 20^o$, $v_T/c = 0.05$, $v_E = eE_o/m_e\omega$)

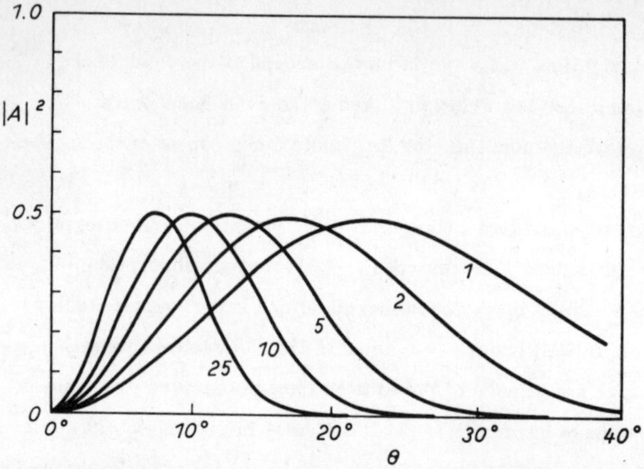

Fig. 4.7 Absorption coefficient versus angle of incidence for different gradient scale lengths at incidence of a p-polarized wave upon a linear density profile (resonance absorption); $v_T/c = 0.05$; curve parameter: L/λ_o

field components, preferentially of $|E_x|$, caused by the superposition of longitudinal perturbations with shorter wavelengths. The spatial amplitude profile of the space charge density (Fig. 4.6 c) indicates the region where the longitudinal component is significant. Its cut-off in direction of lower density is caused by increasing Landau damping, included via the phenomenological damping rate (4.15). It prevents longitudinal perturbations from penetrating the fairly undercritical portion of the density profile.

In Fig. 4.7 the dependence of the absorption coefficient on the angle of incidence is plotted for different gradient scale lengths as a result of numerical integrations of the system (4.31), (4.32). The curves indicate the existence of an angle of incidence with maximum absorption for any gradient scale length. As may be seen, this angle shifts toward smaller values when the density gradient becomes weaker. The optimum angle can be approximated by the relation

$$\sin \theta_{opt} \propto 0.7 \, (k_o L)^{-1/3} \tag{4.33}$$

which was first given in /5/ as a result of an analytical approach. The maximum absorption keeps a level of 50 % independently of the density gradient scale length. As a consequence of the absorption the plasma may be heated through the subsequent transformation of the Langmuir wave energy into thermal plasma energy by collective and/or collisional dissipative effects.

The existence of an angle of incidence associated with maximum absorption can be understood as follows: For small angles of incidence, the electron motion is almost perpendicular to the density gradient and the electric charge separation is small. For large angles of incidence, there exists a sufficiently strong field component along the density gradient; the turning point of the wave and the electrostatic resonance region $N \lesssim N_c$, however, are widely separated. The wave is strongly attenuated in the evanescence region before encountering the critical density and no appreciable tunneling of the radiation occurs. These facts combine to bring about the existence of the optimum angle for absorption.

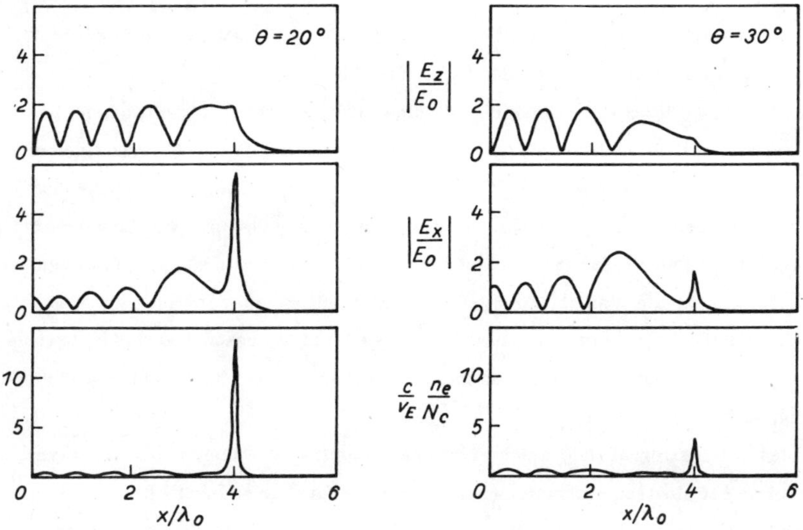

Fig. 4.8 The same as Fig. 4.7; here in the limit of a cold collisional (ν_{ei}/ω = 0.01) plasma

Additional insight in the phenomenon of resonance absorption is provided by going back to the approximation of a cold plasma. Although the longitudinal perturbations then lose their wave character and degenerate to pure space charge oscillations, the mechanism of resonance absorption still works. The equations (4.31), (4.32) simplify in the limit $v_T \to 0$ to

$$-c^2 ik_o \sin\theta \frac{dE_z}{dx} + \omega^2 (\mathcal{E} - \sin^2\theta) E_x = 0, \tag{4.34}$$

$$c^2 \frac{d^2E_z}{dx^2} - ik_o c^2 \sin\theta \frac{dE_x}{dx} + \omega^2 \mathcal{E} E_z = 0. \tag{4.35}$$

After differentiating Eq. (4.34) with respect to x and combinig it with Eq. (4.35) one gets for $\text{div}\,\vec{E} = dE_x/dx + ik_o \sin\theta\, E_z$ the expression

$$\text{div}\,\vec{E} = -\frac{1}{\mathcal{E}}\frac{d}{dx} E_x = \frac{dN/dx}{N_c - N(1 - \nu_{ei}/\omega)} E_x. \tag{4.36}$$

which explicitely demonstrates the existence of a space charge separation as long as the electric field of the wave has a component along the density gradient.

Fig. 4.8 presents solutions of Eqs. (4.34), (4.35) for two angles of incidence ($L/\lambda_o = 4$, $\nu_{ei}/\omega = 0.01$). As expected, an appreciable space charge separation occurs only around the critical layer. The figure illustrates the arguments concerning the existence of an optimum angle for absorption, showing that the longitudinal field peak is less well developed for the higher angle of incidence due to a greater tunneling distance.

Although the spatial field amplitude pattern for a cold and a warm plasma differ significantly, the variation of the absorption with the angle of incidence is practically not altered in the limit $v_T \to 0$. Thus, Fig. 4.7 is representative also for a cold plasma.

According to Eq. (4.36) the space charge density amplitude (and the longitudinal field) become singular in a cold plasma without collisions. The reason for this is that the frequency of the longitudinal component E_x matches there the natural oscillation frequency ω_p of the electrons and the energy fed into the longitudinal oscillations can neither dissipate nor be carried away from the resonance point; thus, the electric field cannot reach a saturated level. A number of different physical effects can limit the amplitude. Besides collisions, a finite background temperature stabilizes the field by convection of the energy out of the resonance region through longitudinal waves with finite group velocity. In this case a final electric field level is established as seen in Fig. 4.6.

For high values of the external field which create high amplitudes of the longitudinal waves, nonlinear effects like hot electron production have to be expected as another stabilizing mechanism for the electric field. This has been seen in particle simulation studies of resonance absorption /9, 17 -19, 25, 27, 28/ as well as in experiments /29, 36, 37/. Although this effect is not included in the fluid model, hot electron production can be assessed via the effect of wavebreaking in a non-stationary fluid approach /23/. This is based on the suggestion that the onset of wavebreaking in the fluid theory signals the creation of hot electrons.

In summary, resonance absorption can be characterized as energy loss of a transverse wave due to its partial transformation into a longitudinal wave by linear mode conversion in a non-uniform plasma. The two distinctive properties of resonance absorption are (a) the dependence on the polarization of the incident wave and (b) the dependence on the angle of incidence. Both dependencies have been experimentally confirmed several times /32 - 36/. Fig. 4.9 presents measurements of Balmer et al. /33/.

The inverse problem, namely the determination of the density scale length in a laser-produced plasma using resonance absorption, has been studied in /38/.

Concerning details of an analytical treatment of resonance absorption we refer to the study /14/ and literature quoted therein where emphasis is given also to mathematical aspects of the problem.

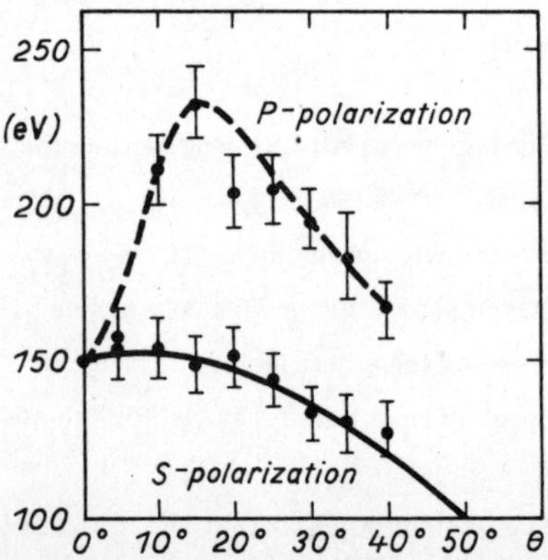

Fig. 4.9 Measured absorption as function of the incidence angle θ for s- and p-polarization after /33/. The electron temperature serves as indicator for the absorption

References

/1/ Budden, K.G.; Radion Waves in the Ionosphere. Cambridge Univ., Cambridge 1961.

/2/ Ginzburg, W.L.; Propagation of Electromagnetic Waves in Plasmas. Pergamon Press, New York 1964.

/3/ White, R.B., Chen, F.F.; Plasma Phys. 16 (1974) 565.

/4/ Forslund, D.W., Kindel, J.M., Lee, K., Lindman, E.L., Morse, R.L.; Phys. Rev. A11 , 679(1975).

/5/ Denisov, N.G.; Zh. Eksp. Teor. Fiz. 31 , 609(1956).

/6/ Piliya, A.D.; Zh. Eksp. Teor. Fiz. 36 , 818(1966).

/7/ Godwin, R.P.; Phys. Rev. Lett. 28 , 85(1972).

/8/ Mueller, M.M.; Phys. Rev. Lett. 30 , 582(1973).

/9/ Freidberg, J.P., Mitchel, R.W., Morse, R.L., Rudsinski, L.L.; Phys. Rev. Lett. 28 , 795(1972).

/10/ Speziale, T., Catto, P.J.; Phys. Fluids 20 , 990(1977).

/11/ Pert, G.J.; J. Plasma Phys. 20 , 175(1978).

/12/ Kull, H.; PLF-Report 16 , Garching 1979.

/13/ Kull, H.J.; Phys. Fluids 26 , 1881(1983).

/14/ Kamp, L.P.J., Weenink, M.P.H.; Physica 122C , 89(1983).

/15/ Koch, P., Albritton, J.; Phys. Rev. Lett. 32 , 1420(1974).

/16/ DeNeef, C.P., DeGroot, J.S.; Phys. Fluids 20 , 1074(1977).

/17/ Albritton, J., Koch, P.; Phys. Fluids 18 , 1136(1975).

/18/ Estabrook, K.G., Valeo, E.J., Kruer, W.L.; Phys. Fluids 18 , 1151(1975).

/19/ DeGroot, J.S., Tull, J.E.; Phys. Fluids 18 , 672(1975).

/20/ Speziale, T., Catto, P.J.; Phys. Fluids 22 , 681(1979).

/21/ Kruer, W.L.; Phys. Fluids 25 , 2324(1982).

/22/ Adam, J.C., Serveniere, A.G., Laval, G.L.; Phys. Fluids 25 , 376(1982).

/23/ Bezzerides, B., Gitomer, S.J.; Phys. Fluids 26 , 1359(1983).

/24/ Estabrook, K.G., Valeo, E., Kruer, W.L.; Phys. Lett. 49A , 109(1974).

/25/ Forslund, D.W., Kindel, J.M., Lee, K.; Phys. Rev. Lett. 39 , 284(1977).

/26/ Estabrook, K., Kruer, W.L.; Phys. Rev. Lett. 40 , 42(1978).

/27/ Bezzerides, B., Gitomer, S.J., Forslund, D.W.; Phys. Rev. Lett. 44 , 651(1980).

/28/ Bezzerides, B., Gitomer, S.J.; Phys. Rev. Lett. 46 , 593(1981).

/29/ Stenzel, R.L., Wong, A.Y., Kim, H.C.; Phys. Rev. Lett. 32 , 654(1974).

/30/ DeNeef, P.; Phys. Rev. Lett. 39 , 997(1977).

/31/ Mizuno, K., DeGroot, J.S., Kehl, F.; Phys. Rev. Lett. 49 , 1004(1982).

/32/ Pearlman, J.S., Thomson, J.J., Max, C.E.; Phys. Rev. Lett. 38 , 1397(1977).

/33/ Balmer, J., Donaldson, T.P.; Phys. Rev. Lett. 39 , 184(1977).

/34/ Manes, K., Rupert, V., Auerbach, J., Lee, P., Swain, J.; Phys. Rev. Lett. 39 , 281(1977).

/35/ Pearlman, J.S., Matzen, M.K.; Phys. Rev. Lett. 39, 140(1977).

/36/ Godwin, R., Sachsenmaier, P., Sigel, R.; Phys. Rev. Lett. 39, 1198(1977).

/37/ Kolodner, P., Yablonovitch, E.; Phys. Fluids 24, 759(1981).
dto, Phys. Rev. Lett. 37, 1754(1976).
dto, Phys. Rev. Lett. 45, 1790(1980).

/38/ Luther-Davies, B.; Opt. Comm. 37, 197(1981).

5 Structure resonances

As discussed in the preceding section, strongly enhanced absorption ($|A| \leq 50\ \%$) occurs at the incidence of a p-polarized wave on an inhomogeneous plasma when the angle of incidence is appropriately chosen. The underlying basic mechanism is the linear transformation of transverse waves into longitudinal waves in nonuniform plasmas near the critical layer $\varepsilon(x) = 0$.

Another kind of resonances arises due to coherence effects involving plasma eigenmodes whose existence is based upon particular spatial structures of the plasma density. These resonances, called structure resonances, are determined by global properties of the plasma density profile, in contrast to the local resonance at $\varepsilon(x) = 0$, studied in the preceding section. For s-polarized waves the investigation of structure resonances in plasmas is closely related to well-known problems of classical optics in layered media /1, 2/. The finite plasma layer and the two-step density profile consisting of an underdense plasma layer bounded by an overdense one are the most simple examples. In regard to self-consistent situations at non-linear wave-plasma interaction, inhomogeneous density profiles containing a cavity near the critical surface have received some attention /3 - 5/. For modeling this caviton resonance, a three-step density profile is used where two overdense layers are separated by an underdense one. For appropriate widths of the cavity, this systems acts as a resonator leading to strong absorption even for weakly collisional plasmas. For p-polarized waves, additional possibilities of interferences are originated by surface waves occuring in plasmas with steep density gradients. Under definite conditions concerning the density profile and/or the angle of incidence, these waves are excited resonantly by the incident electromagnetic wave, indicated by an enhancement of the electric field of the surface wave within the plasma. As a consequence, large absorption of the incoming energy may arise in the presence of weak damping.

The structure resonances mentioned above are described by the cold-plasma model which neglects finite electron temperature effects. Obviously, for s-polarized waves this approximation does not introduce any restriction. In the case of p-polarized waves, however, the excitation of longitudinal waves in a plasma of finite temperature may alter the properties of the cold-plasma surface-wave resonances. A second and more important aspect concerns structure resonances generated by purely longitudinal waves which appear only in plasmas of finite temperature. In this connection, the caviton resonance is of particular interest because of its relevance to self-consistent density profiles formed by the ponderomotive force. During resonance, plasma waves are trapped in the density cavity and driven to a strongly enhanced amplitude. The resulting absorption may exceed the 50 % limit of resonance absorption at a linear density profile.

5.1 Resonances at s-polarization

In this section we shall examine electromagnetic structur resonances of s-polarized waves. For simplicity, we consider parallel propagation with the wave vector parallel to the x-axis. In a straightforward way, all results can easily be transferred to obliquely propagating s-polarized waves by simply replacing $\mathcal{E} = 1 - N/N_c$ in the wave equation through $\mathcal{E} - \sin^2\theta$ where θ is the angle of incidence (cf. Eq. 4.30).

We start with the simple case of the propagation of an electromagnetic wave through a homogeneous underdense plasma slab (Fig. 5.1). Here, main properties of electromagnetic structure resonances become already evident. Furthermore, the analysis used here needs only slight extension in order to handle the more complex density models.

For the three regions 0 ($x < 0$), I ($0 \le x \le d$) and II ($x > d$) the stationary wave equation has the following solutions

$$\begin{aligned} &0: && E_o(x) = E_o^+ e^{ik_o x} + E_o^- e^{-ik_o x}, \\ &\text{I}: && E_1(x) = E_1^+ e^{ik_1 x} + E_1^- e^{-ik_1 x}, \\ &\text{II}: && E_2(x) = E_2^+ e^{ik_o (x-d)} \end{aligned} \tag{5.1}$$

with

$$k_o = \omega/c, \quad k_1 = k_o\sqrt{\mathcal{E}},$$

where ω is the frequency of the incident wave.

Dissipation effects are neglected for the present purpose. In the regions 0 and I (plasma slab) the electric field results from the superposition of the incoming and reflected wave, whereas in the vacuum region II only the transmitted wave exists. The complex amplitudes E_o^-, E_1^+, E_1^- and E_2^+ are determinded by the matching conditions for both the electric and magnetic field at the boundaries $x = 0$ and $x = d$:

$$E_o^+ + E_o^- = E_1^+ + E_1^-$$

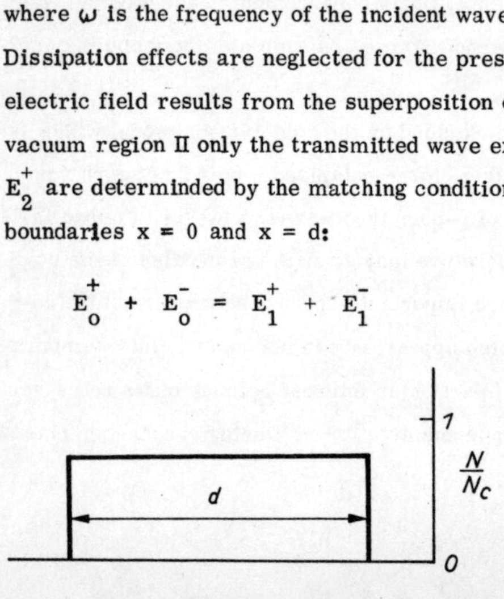

Fig. 5.1 Plasma slab of the density N and of the length d

$$k_o (E_o^+ - E_o^-) = k_1 (E_1^+ - E_1^-)$$

$$E_1^+ e^{ik_1 d} + E_1^- e^{-ik_1 d} = E_2^+$$

$$k_1 \left(E_1^+ e^{ik_1 d} - E_1^- e^{-ik_1 d} \right) = k_o E_2^+ \tag{5.2}$$

$E_o^+ = E_o$ is the amplitude of the incident wave. The reflection coefficient R and the transmission coefficient T (both complex) are determined by the relations

$$R = E_o^- / E_o , \quad T = E_2^+ / E_o .$$

Solving the linear system of Equations (5.2), the well-known result

$$|R|^2 = \frac{(\varepsilon - 1)^2 \tan^2 k_1 d}{(\varepsilon + 1)^2 \tan^2 k_1 d + 4\varepsilon} \tag{5.3}$$

is obtained.

This expression shows that the reflection coefficient oscillates with the length of the layer d and special states having $|R|^2 = 0$ equivalent to total transmission occur at certain lengths $d^{(m)}$, given by

$$k_1 d^{(m)} = m\pi , \quad m = 1, 2, 3, \ldots \tag{5.4}$$

Introducing the wave length of the electromagnetic wave within the plasma slab, $\lambda_1 = 2\pi / (k_o \sqrt{\varepsilon})$, the relation (5.4) obtains a simple geometrical interpretation expressing that at total transmission the length of the layer fits multiples of half a wave length λ_1,

$$d^{(m)} = m \lambda_1 / 2 . \tag{5.5}$$

This situation represents a resonance state which is characterized by a maximum electric field within the plasma slab. Using the expressions for E_1^+, E_1^-

$$\frac{E_1^+}{E_o} = \frac{k_1 + k_o}{2 k_1} , \quad \frac{E_1^-}{E_o} = \frac{k_1 - k_o}{2 k_1} ,$$

which follow from Eqs. (5.2) after setting $E_o^- = 0$, the spatial structure of the plasma field takes the form

$$\frac{|E_1(x)|}{E_o} = \frac{1}{k_1} \left[k_1^{\,2} \cos^2 k_1 x + k_o^{\,2} \sin^2 k_1 x \right]^{1/2} ,$$

and it is easily seen that the maximum electric field is given by

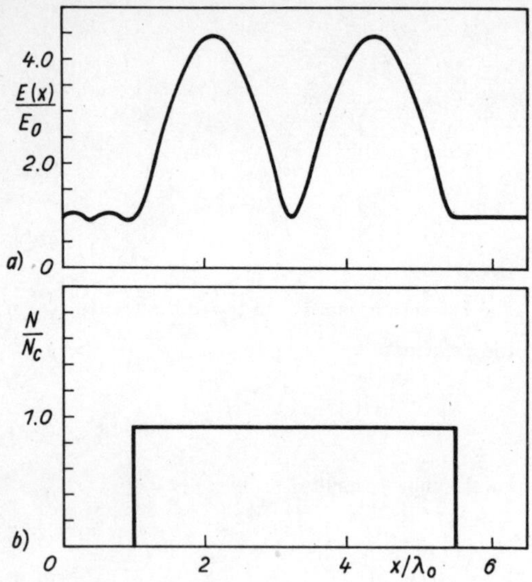

Fig. 5.2 Spatial amplitude profile (a) at resonant propagation of an electromagnetic wave through a plasma slab (b); $N/N_c = 0.95$, $d \simeq 4.5\ \lambda_o$

$$\frac{|E_1|_{max}}{E_o} = \frac{k_o}{k_1} = \frac{1}{\sqrt{\varepsilon}} \tag{5.6}$$

That means, a significant enhancement of the electric field appears only for plasma densities near to the critical density, that is $\varepsilon \ll 1$.

As an example we consider a plasma slab with the density $N/N_c = 0.95$. Fig. 5.2 shows the spatial structure of the electric field amplitude at resonance realized by choice of the slab length d according to $d = d^{(2)} = \lambda_1 \simeq 4.5\ \lambda_o$. The standing wave structure results from the coherent superposition of the forward and backward travelling wave within the plasma layer. Consistent with formula (5.6) the field amplitude reaches a maximum of about 4.5 times the vacuum value.

Up to now, our considerations have been confined to collisionless plasmas. If we include dissipative effects by a collision frequency ν, the expression for the absorption coefficient

$$|A|^2 = \frac{\nu}{c} \int_o^d \frac{|E_1(x)|^2}{E_o^2}\, dx \tag{5.7}$$

indicates that the field enhancement at resonance has serious consequences to the absorption of electromagnetic energy in the plasma. In the approximation of weak collisions ($\mathrm{Im}\, k_1 d \ll 1$), it is easily shown that at resonance an optimum absorption of 50 % can be reached. Because of its importance, the problem of absorption will be considered later in more detail.

Similar structure resonances as described above also occur when an overdense plasma ($N_2/N_c > 1$) is adjacent to the underdense plasma slab ($N_1/N_c < 1$) at the right-hand side (Fig. 5.3). Resonances of such a two-step density profile and their consequences for electromagnetic wave absorption have been discussed by Mayer et al. /6/.

The calculation of the electric field can be carried out in full analogy to the previous case with

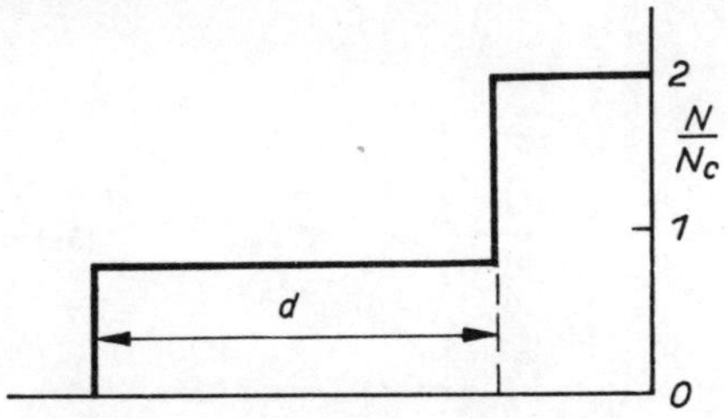

Fig. 5.3 Two-step density profile consisting of an underdense plasma slab ($N_1 < N_c$) bounded by an overdense plasma ($N_2 > N_c$)

the only difference that there is an evanescent penetration of the wave into the overdense region II. Corresponding to the densities N_1 and N_2, the wave numbers k_1 and k_2 are given by

$$k_1 = k_o \sqrt{\varepsilon_1} \, , \quad k_2 = k_o \sqrt{\varepsilon_2} \tag{5.8}$$

with

$$\varepsilon_j = 1 - \frac{N_j/N_c}{1 + i\, \nu_j/\omega} \, , \quad j = 1, 2$$

where the collision frequencies ν_1, ν_2 are included to describe energy dissipation. Using the modified version of the matching conditions (5.2), the wave amplitudes are given by

$$R = \frac{E_o^-}{E_o} = \left[(k_o + k_1)(k_1 - k_2)\, e^{ik_1 d} + (k_o - k_1)(k_1 + k_2)\, e^{-ik_1 d} \right] / D$$

$$\frac{E_1^+}{E_o} = 2 k_o (k_1 + k_2)\, e^{-ik_1 d} / D$$

$$\frac{E_1^-}{E_o} = 2 k_o (k_1 - k_2)\, e^{ik_1 d} / D$$

$$\frac{E_2^+}{E_o} = 4 k_o k_1 / D \tag{5.9}$$

with

$$D = (k_o - k_1)(k_1 - k_2)\, e^{ik_1 d} + (k_o + k_1)(k_1 + k_2)\, e^{-ik_1 d} \, . \tag{5.10}$$

It is convenient to introduce the following dimensionless quantities

$$\alpha = \frac{k_1}{k_o} = \sqrt{\varepsilon_1} \, , \quad \beta = i \frac{k_1}{k_2} \, , \quad \delta = k_1 d \, ,$$

$$G(\beta, \delta) = \frac{1 - i\beta}{1 + i\beta}\, e^{-2i\delta} \tag{5.11}$$

$$S(\alpha, \beta, \delta) = (1 - \alpha) - (1 + \alpha)\, G(\beta, \delta) \, .$$

In this notation we obtain

$$R = \frac{S(-\alpha,\beta,\delta)}{S(\alpha,\beta,\delta)}, \quad \frac{E_1^+}{E_o} = -2\frac{G(\beta,\delta)}{S(\alpha,\beta,\delta)}, \quad \frac{E_1^-}{E_o} = \frac{2}{S(\alpha,\beta,\delta)} . \tag{5.12}$$

Considering the collisionless case ($\nu_1 = \nu_2 = 0$), then α and β are real quantities, which leads to

$$|S(\alpha,\beta,\delta)| = |S(-\alpha,\beta,\delta)|$$

and, consequently, $|R|^2 = 1$. This is not surprising because the overdense plasma acts as a total reflector.

Looking for structure resonances of the two-step profile, however, the behaviour is only slightly modified compared with the simple plasma layer. According to Eq. (5.12), the electric field within the underdense layer becomes an optimum where $|S(\alpha,\beta,\delta)|$ becomes a minimum. This leads to the condition

$$G(\beta,\delta) = 1$$

which is equivalent to

$$\exp(-i\phi) = 1 ,$$

where ϕ is given by

$$\phi = 2\delta + 2 \arctan \beta .$$

Thus, the resonance condition takes the form

$$k_1 d = m\pi - \arctan(ik_1/k_2) , \quad m = 1, 2, \ldots \tag{5.13}$$

and differs from the corresponding relation (5.4) of the plasma slab only by the additional arc tan term.

In the resonance case the amplitudes E_1^+ and E_1^- are given by

$$E_1^+ = -E_1^- = \frac{E_o}{\sqrt{\varepsilon_1}} .$$

As result, the electric field within the underdense plasma layer can be written as

$$E_1(x) = \frac{2iE_o}{\sqrt{\varepsilon_1}} \sin(k_o\sqrt{\varepsilon_1}\,x) . \tag{5.14}$$

Thus, the maximum field enhancement at resonance is given by

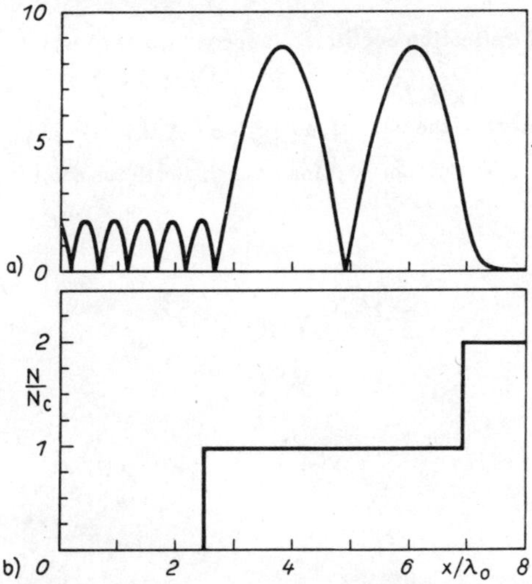

Fig. 5.4 Spatial amplitude profile (a) belonging to the resonant two-step density profile (b); $N_1/N_c = 0.95$, $N_1 = N_3 = 2.0$, $d = 4.5\ \lambda_o$, $\nu_1 = \nu_2 = 0$

$$\frac{|E_1|\ \mathrm{max}}{E_o} = 2/\sqrt{\varepsilon_1}\ . \qquad (5.15)$$

The factor two in this expression, which is missing in Eq. (5.6) for the simple layer, results from the total reflection of the wave at the overdense plasma. An example for the spatial field structure at resonance is shown in Fig. 5.4.

Coming back to real dissipative plasmas, the close correlation between structure resonance and electromagnetic energy absorption becomes evident from Fig. 5.5 where the field enhancement is shown together with the reflection coefficient as a function of the underdense plateau length d for two values of the collision frequency $\nu = \nu_1 = \nu_2$. It is clearly seen that the peaks of the maximum electric field amplitude at resonance are correlated with a remarkable

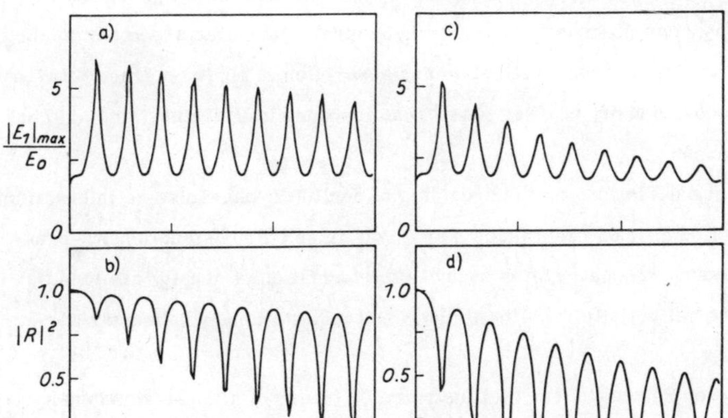

Fig. 5.5 Field enhancement (a, c) and reflection coefficient (b, d) versus the plateau length d of a two-step density profile for two collision frequencies; $\nu_1/\omega = 10^{-3}$ (a, b), $\nu_1/\omega = 5.10^{-3}$ (c, d), $N_1/N_c = 0.9$, $N_2/N_c = 2$

reduction of the reflectivity. At certain lengths the reflection coefficient approaches zero which corresponds to total absorption.

According to Eq. (5.12), $|R|^2 = 0$ is equivalent to the conditions $|S(-\alpha, \beta, \delta)| = 0$, and in the approximation of weak collisions ($\mathrm{Im}\, k_1 d \ll 1$), one obtains as additional condition for total absorption:

$$\mathrm{Im}\,(k_1 d) = \sqrt{\varepsilon_1}.$$

With

$$\mathrm{Im}\,(k_1) = \frac{k_o}{2\sqrt{\varepsilon_1}}\, \nu_1/\omega$$

it follows

$$k_o d\ \nu_1/\omega = 2\,\varepsilon_1 .$$

Eliminating the length d by use of the approximated resonance condition $k_1 d = m$ (m = 1, 2, 3, ...) yields

$$m\pi\ \nu_1/\omega = 2(\varepsilon_1)^{3/2}. \tag{5.16}$$

Taking the parameters of Fig. 5.5 (c, d) - $\varepsilon_1 = 0.1$, $\nu/\omega = 5.10^{-3}$ - total absorption is approximately reached at m = 4. This can be seen in Fig. 5.5 (d).

So far, we have considered structure resonances which are caused by the interference of waves propagating in an underdense plasma and where a remarkable electric field enhancement requires the plasma density to be near to the critical one, i. e. $N \approx N_c$ or $\varepsilon \ll 1$.

Another kind of structure resonance appears in plasmas whose density profile contains a cavity in which electric energy can resonantly be trapped. Several interesting aspects of such a configuration with regard to "anomalous" reflection and absorption of electromagnetic waves in plasmas with non-monotonous density profiles have been discussed in literature since 1977 /3, 7, 8/.

At the same time, numerical calculations on the dynamics of nonlinear wave-plasma interaction have shown that these so-called caviton resonances play a role in self-consistent density-profile modifications. When passing resonant states owing to the formation of density cavitons by the ponderomotive force, the reflectivity exhibits a significant non-stationary behaviour /4, 5, 9 - 11/.

Main features of the caviton resonance can be illustrated by means of a three-step density profile (Fig. 5.6) where an underdense plasma layer ($N_2 < N_c$) is imbedded by two adjacent overdense plasma layers (N_1, $N_3 > N_c$). It should be pointed out, that the considered configuration has much similarity to a metal interference filter where a $\lambda/4$-sheet is surrounded by two thin metal layers. There, resonance is indicated by maximum transmission.

For a better insight in the nature of the caviton resonance, we consider at first an under-

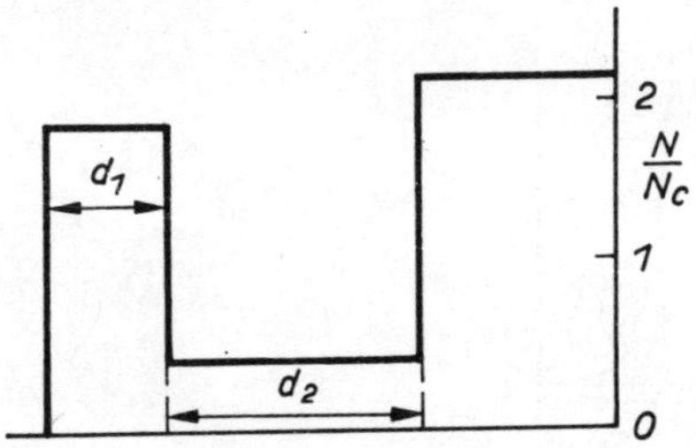

Fig. 5.6 Caviton density-profile with $N_1 > N_c$, $N_2 < N_c$ and $N_3 > N_c$

dense cavity within an unbounded overdense plasma and look for eigensolutions of the electric field with $E_1(x) \longrightarrow 0$ for $x \longrightarrow -\infty$ and $E_3(x) \longrightarrow 0$ for $x \longrightarrow +\infty$, where E_1 and E_3 are the fields in the overdense regions I and III, resp. Matching the boundary conditions leads to a homogeneous, algebraic system for the field amplitudes analogously to Eq. (5.2), and this system has a non-trivial solution only when the condition

$$\tan(k_2 d_2) = -i \frac{k_1 k_2 + k_2 k_3}{k_2^2 + k_1 k_3} \tag{5.17}$$

is fulfilled, where $k_i = k_o \sqrt{\varepsilon_i}$ $(i = 1, 2, 3)$. Because the wave numbers k_1 and k_3 (in the absence of collisions) are purely imaginary, the following notations are introduced

$$\varkappa_1 = -i \frac{k_1}{k_o} = \sqrt{N_1/N_c - 1}\,, \quad \varkappa_3 = -i \frac{k_3}{k_o} = \sqrt{N_3/N_c - 1}\,. \tag{5.18}$$

For a given density profile (N_1, N_2, N_3, d_2) the relation (5.17) represents a transcendental equation determining the critical density N_c and, thus, the eigenfrequency of the caviton profile. However, it is more convenient to determine the resonant caviton width d_2 if the normalized densities N_i/N_c $(i = 1, 2, 3)$ are given. Then, one obtains from Eq. (5.17) the simple resonance condition

$$k_2 d_2^{(m)} = \arctan \frac{\varkappa_1}{\sqrt{\varepsilon_2}} + \arctan \frac{\varkappa_3}{\sqrt{\varepsilon_2}} + (m-1)\pi\,, \quad m = 1, 2, 3, \ldots \tag{5.19}$$

This relation becomes particularly simple in the case

$$\varkappa_1 = \varkappa_3 = \varepsilon_2 = 1\,, \tag{5.20}$$

that is,

$$N_1 = N_3 = 2 N_c\,, \quad N_2 = 0\,. \tag{5.21}$$

In this special case, the resonance is given by

$$k_o d_2^{(m)} = (2m-1)\pi/2 \quad \text{or} \quad d_2^{(m)} = (2m-1)\lambda_o/4\,, \tag{5.22}$$

where λ_o is the vacuum wave length. That means, resonances occur, if uneven multiples of

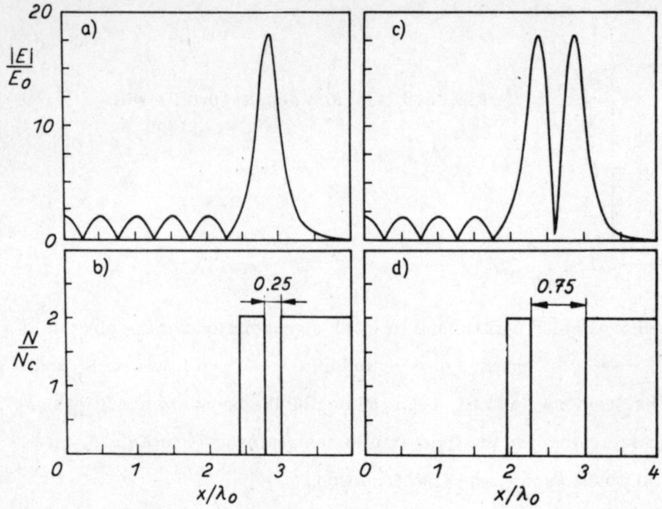

Fig. 5.7 Spatial amplitude profiles (a, c) which belong to the resonant caviton density-profiles (b, d). The caviton width is $d_2 = 0.25\ \lambda_o$ (b) and $d_2 = 0.75\ \lambda_o$ (d), resp.; $N_1/N_c = N_3/N_c = 2.0$, $N_2 = 0$, $d_1 = 0.35\ \lambda_o$, $\nu_1 = \nu_3 = 0$

$\lambda_o/4$ match the cavity width.

Now we come back to the more realistic caviton profile with a finite width d_1 of the overdense plasma barrier at the left-hand side. An interesting problem concerns the excitation of the caviton resonance discussed above by an incident electromagnetic wave and the question how the resonant field enhancement within the cavity depends on the parameters (density, length) of the supercritical layer. In order to calculate the spatial structure of the electric field, the usual procedure is used for solving the algebraic systems for the six amplitudes E_o^-, E_1^+, E_1^-, E_2^+, E_2^- and E_3^+, which follows from the matching conditions at the boundaries $x = 0$, $x = d_1$, $x = d_1 + d_2$.

Results are shown in Fig. 5.7 for the first ($m = 1$) and second resonance ($m = 2$) using the special density profile (5.21). The width of the plasma barrier is taken as $d_1 = 0.35\ \lambda_o$, collisions are neglected. It is obvious again, that without dissipation all the energy is reflected by the overdense plasma. But note the strong enhancement of the electric field within the cavity.

Taking into account weak collisions ($\nu_{1,3}/\omega \ll 1$), then the following expressions for the absorption coefficient and the caviton field maximum can be derived:

$$|A|^2 = \frac{4\,G\,G_c}{\tilde{D}^2}\,, \qquad \frac{|E_2|_{max}}{E_o} = \frac{2\sqrt{G_c}}{\tilde{D}}\, e^{\varkappa_1 k_o d_1} \tag{5.23}$$

where

$$G = \frac{1}{2}\left[\frac{\nu_1/\omega}{\varkappa_1} + \frac{\nu_3/\omega}{\varkappa_3}\right] e^{2\varkappa_1 k_o d_1}\,,$$

$$\tilde{D} = (H - H_c)^2 + (G - G_c)^2\,, \quad H = k_o\,(d_2 - d_2^{(m)})\, e^{2\varkappa_1 k_o d_1}\,,$$

$$H_c = \frac{\tilde{\alpha}}{1+\tilde{\alpha}^2}, \quad G_c = \frac{1}{1+\tilde{\alpha}^2}, \quad \tilde{\alpha} = \frac{1-x_1^2}{2x_1}. \tag{5.24}$$

Considering the expressions for $|A|^2$ and $\tilde{D}$, it is immediately seen that total absorption can be realized when the conditions

$$H = H_c, \qquad G = G_c \tag{5.25}$$

are fulfilled simultaneously by an appropriate choice of the parameters. Apart from a small shift, the first condition coincides with the resonance condition (5.19), $d_2 = d_2^{(m)}$. For given densities and a fixed collision frequency, the second condition can be satisfied by variation of the width d_1 of the overdense plasma barrier.

Since $\tilde{D} = 2\,G_c$, the electric field enhancement at total absorption is given by

$$\frac{|E_2|_{max}}{E_o} = \frac{1}{\sqrt{G_c}} \exp(\chi_1 k_o d_1) = \frac{1+\chi_1^2}{2\chi_1} \exp(\chi_1 k_o d_1). \tag{5.26}$$

In Fig. 5.8, $|A|^2$ and $|E_2|_{max}/E_o$ are shown as a function of d_1 for two collision frequencies. It is evident, that the enhancement of the electric field does not grow monotonously with increasing d_1; it reaches its maximum value at $|A|^2 = 1$.

In the collisionless cas ($G = 0$), however, there is no limitation, and because of $\tilde{D} = G_c$, the field enhancement at that d_1 which yields total absorption for finite collision frequency is two times larger. For the parameters of Fig. 5.7 ($\chi_1 = 1.0$, $d_1 = 0.35\,\lambda_o$), from

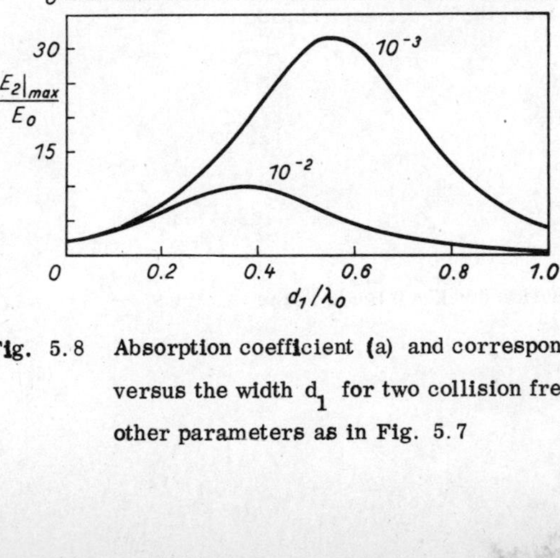

Fig. 5.8 Absorption coefficient (a) and corresponding field enhancement (b) at resonance versus the width d_1 for two collision frequencies; $\nu_1/\omega = \nu_3/\omega = 10^{-2}, 10^{-3}$, other parameters as in Fig. 5.7

$$\frac{E_{2\,max}}{E_o} = \frac{1 + \chi_1^2}{\chi_1} \exp(\chi_1 k_o d_1) \tag{5.27}$$

one gets

$$|E_2|_{max}/E_o \simeq 18 .$$

5.2. Surface wave resonances

In our considerations in Section 5.1 we have assumed that the electromagnetic wave propagates parallel to the density gradient. As mentioned already, these results are representative also for obliquely propagating waves as long as no electric field component exists in the direction of the density gradient (s-polarisation). For p-polarized waves and oblique incidence, however, electron density perturbations are generated in the regions where the plasma density changes abruptly. This ist connected with the excitation of surface waves propagating along the density jump. Therefore, new kinds of structure resonances are to be expected under these conditions.

The analysis of structure resonances associated with surface waves is similar to that developed in Section 5.1, one only has to take into account the additional (parallel) field component in the matching conditions. Considering a homogeneous plasma layer, the electric field polarized in the x, z plane can be written as

$$E_x(x, z) = E_x^+ e^{i(k_x x + k_z z)} + E_x^- e^{i(-k_x x + k_z z)} \tag{5.28}$$

$$E_z(x, z) = E_z^+ e^{i(k_x x + k_z z)} + E_z^- e^{i(-k_x x + k_z z)} \tag{5.29}$$

with

$$k_z = k_o \sin\theta , \tag{5.30}$$

where θ is the angle of incidence.
The x-component of the wave vector follows from the dispersion relation

$$k_x^2 + k_z^2 = k_o^2 \mathcal{E} , \tag{5.31}$$

that is,

$$k_x = k_o \sqrt{\mathcal{E} - \sin^2\theta} . \tag{5.32}$$

Using the expressions (5.28), (5.29), the relation div $\vec{E} = 0$ leads to the relations

$$k_x E_x^+ + k_z E_z^+ = 0 , \qquad -k_x E_x^- + k_z E_z^- = 0 , \tag{5.33}$$

and thus, the polarization of both the forward and backward travelling waves is given by

$$E_x^+ = -p\,E_z^+ , \qquad E_x^- = p\,E_z^- , \tag{5.34}$$

where

$$p = \frac{k_z}{k_x} = \frac{\sin\theta}{\sqrt{\mathcal{E} - \sin^2\theta}} \tag{5.35}$$

is the polarization factor.

Then, Eq. (5.28) takes the form

$$E_x(x, z) = p\left[-E_z^+ e^{ik_x x} + E_z^- e^{-ik_x x}\right] e^{ik_z z} . \tag{5.36}$$

To formulate the matching conditions, the y-component of the magnetic field

$$B_y(x, z) = \frac{i}{k_o}\left(\frac{\partial}{\partial x} E_z - \frac{\partial}{\partial z} E_x\right) \tag{5.37}$$

has to be calculated. Using Eq. (5.36), we find

$$B_y(x, z) = i\frac{K}{k_o}\left(E_z^+ e^{ik_x x} - E_z^- e^{-ik_x x}\right) e^{ik_z z} \tag{5.38}$$

with

$$K = k_o \frac{\mathcal{E}}{\sqrt{\mathcal{E} - \sin^2\theta}} . \tag{5.39}$$

Being aware of the matching conditions, the expression (5.38) for B_y suggests how the results of the previous section can be applied to the case of p-polarized waves. We have simply to introduce the following substitutions:

$$\begin{aligned} k_o &\longrightarrow K_o = k_o/\cos\theta , \\ k_j &\longrightarrow K_j , \\ \exp(ik_j d) &\longrightarrow \exp(ik_{jx} d), \quad j = 1, 2 . \end{aligned} \tag{5.40}$$

To study resonances occuring at a two-step density profile (see Fig. 5.3), we consider the eigensolutions of the unbounded step-profile having the densities N_1 and N_2 with $\mathcal{E}_1 - \sin^2\theta < 0$ and $N_2 > N_c$. Because of $k_{1x}^2 < 0$ and $k_{2x}^2 < 0$, only evanescent solutions exist. Using the matching conditions at $x = 0$,

$$E_{1z}^- = E_{2z}^+ , \qquad -K_1 E_{1z}^- = K_2 E_{2z}^+ \tag{5.41}$$

with K_i ($i = 1, 2$) from Eq. (5.39), one gets the relation

$$K_1 = -K_2 . \tag{5.42}$$

This is the condition for the existence of a surface wave propagating in z-direction along the density jump. For given ε_1 and θ, Eq. (5.42) yields ε_2 and, hence, N_2/N_c /12/:

$$\varepsilon_1^2 (\varepsilon_2 - \sin^2 \theta) = \varepsilon_2^2 (\varepsilon_1 - \sin^2 \theta) \tag{5.43}$$

or

$$\frac{N_2}{N_c} = 1 + \frac{(1 - N_1/N_c) \sin^2 \theta}{N_1/N_c - \cos^2 \theta} . \tag{5.44}$$

Considering the special case of vacuum in the region I ($N_1 = 0$), in quasistatic approximation $k_z/k \gg 1$ the well-known condition for the existence of surface waves in a plasma half-space

$$N = 2 N_c , \qquad \omega = \omega_p \sqrt{2} \tag{5.45}$$

is obtained. These waves, however, cannot couple with electromagnetic waves in vacuum. For $\theta = 30^{\circ}$ and $N_1/N_c = 0.8$, Eq. (5.44) gives the (resonant) density $N_2/N_c = 2.0$. This particular parameter combination will frequently be used in what follows.

Consider now a two-step density profile. Under resonant conditions ($K_1 = -K_2$) a surface wave propagates along the density jump whereby the electric field amplitude reaches its maximum at x = d and decreases exponentially in x-direction to both sides. An example for the spatial structure of the electric field using the standard parameter set above is shown in Fig. 5.9. At the points where the density has a jump (x = 0, x = d), the x-component of the electric field has a discontinuity resulting from the jump of the polarization factor given by Eq. (5.35).

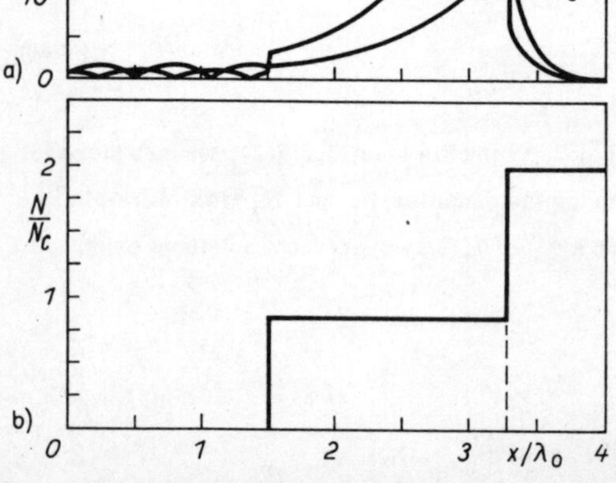

Fig. 5.9 Spatial amplitude profile of the surface wave (a) which belongs to the resonant two-step density profile (b); $\theta = 30^{\circ}$, $N_1/N_c = 0.8$, $N_2/N_c = 2.0$, $\nu_1 = \nu_2 = 0$

The maximum of E_z at $x = d$ is determined by the amplitude E^+_{2z}, and from Eq. (5.9) with $K_1 = -K_2$ one obtains

$$\frac{|E_z|\max}{E_o} = \frac{2K_o}{|K_1 - K_o|}\exp(\varkappa_1 k_o d) = \frac{2\varkappa_1}{\sqrt{\varkappa_1^2 + \varepsilon_1^2}}\exp(\varkappa_1 k_o d), \tag{5.46}$$

where

$$\varkappa_1 = \sin^2\theta - \varepsilon_1 . \tag{5.47}$$

Thus, the maximum field amplitude increases exponentially with increasing plateau length d as long as dissipation effects, neglected here, do not lead to a limitation.

The absorption coefficient can easily be calculated from Eq. (5.9) using the substitutions (5.40),

$$|A|^2 = 1 - |R|^2 \tag{5.48}$$

with

$$R = \left[(K_o + k_1)(K_1 - K_2)e^{-\varkappa_1 k_o d} + (K_o - K_1)(K_1 + K_2)e^{\varkappa_1 k_o d}\right]/D \tag{5.49}$$

and

$$D = (K_o - K_1)(K_1 - K_2)e^{-\varkappa_1 k_o d} + (K_o + K_1)(K_1 + K_2)e^{\varkappa_1 k_o d}. \tag{5.50}$$

For our standard case ($\theta = 30^o$, $N_1/N_c = 0.8$, $N_2/N_c = 2.0$) both the absorption coefficient and the corresponding field maximum as a function of the plateau length d are shown in Fig. 5.10 with ν/ω (10^{-2}, 10^{-3}, 10^{-4}) as parameter.

As far as a stepwise constant density profile is considered, effects like excitation of longitudinal plasma waves via linear mode coupling and resonance absorption are excluded. When

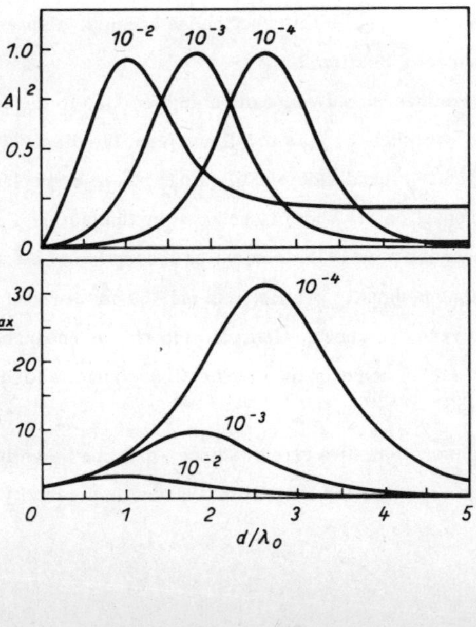

Fig. 5.10 Absorption coefficient (a) and maximum field amplitude $|E_z|/E_o$ (b) at resonance versus the plateau length d for three collision frequencies; $\nu_I/\omega = 10^{-2}, 10^{-3}, 10^{-4}$; other parameters as in Fig. 5.9

assuming finite slopes at the density transition regions, these effects cannot longer be ignored. To study such a situation, a number of calculations have been performed using a plasma density profile where the transition from the lower density plateau to the upper one occurs over a linear ramp with the scale length L. Since it is difficult to achieve a conclusive picture on the parameter dependence of the absorption coefficient, we want to formulate only some tendencies:

- Finite density gradients lead to resonance broadening in such a way, that enhanced absorption occurs over a broader region of angles of incidence compared with the very sharp dependence in the case of discontinuous profiles.
- At density profiles with finite slopes, resonance absorption occurs. This is indicated by the fact, that the optimum plateau length, i. e., the length for maximum absorption, is shifted to smaller values with decreasing density gradient. The same effect has been observed at density profiles containing infinite gradients if the collision frequency increases (Fig. 5.10).

We point finally to Ref. /13/, where a similar approach has been used and the condition $k_o L \mathcal{E}_1 \sin\theta < 0.42$ for enhanced resonance absorption has been given.

5.3 Electrostatic structure resonances

So far, the study of structure resonances was based on the cold plasma model. Whereas a finite electron temperature has only negligible influence on the dispersion of s-polarized waves and thus on the corresponding structure resonances, the propagation of p-polarized waves is changed by the excitation of electron plasma waves (Section 4.3). But, in most cases only modifications of the cold plasma structure resonances arise. Therefore, this point shall not further be discussed here. Our emphasis is directed on electrostatic structure resonances which are based exclusively on coherence effects of electron plasma waves and thus are very sensitive against variations of the electron temperature. In this section we consider a special problem which is related to the caviton resonance of Section 5.1.

The resonance condition (5.19), valid for transverse waves, can be applied also to the case of purely longitudinal waves after replacing c^2 through $3v_T^2$ as it follows from the dispersion relations (4.6), (4.7). Consequently, k_2 on the left-hand side of Eq. (5.19) has to be replaced by $(c/\sqrt{3}\, v_T)\, k_2$, which means, that the resonant cavity width is reduced by the factor $\sqrt{3}\, v_T/c$. The question what role these resonances play in the context of absorption of electromagnetic waves in plasmas having non-monotonous density profiles can only be answered if the conditions for their excitation by the incident wave are known. General criteria for enhanced absorption similar to the "displacement rule" (4.33) at resonance absorption cannot be formulated.

When looking for conditions for optimum absorption, two circumstances must be taken into consideration. On the one hand, the resonance condition for longitudinal waves in the cavity must be satisfied approximately; on the other hand, an effective coupling of the longitudinal waves to the incident transverse field must be guaranted. Hence, optimum absorption requires a compromise between these two conditions. For a given depth of the cavity, for instance, this can be realized by a proper choice of the density barrier in front of the cavity which must both allow a sufficient wave tunneling and prevent the excited plasma waves from leaving the cavity.

The operation of electrostatic structure resonances can be demonstrated by use of an appropriately chosen continuous caviton density profile. Because the wave length of plasma waves is a function of the electron temperature, the resonances can be tuned by variation of this parameter. Then, they are indicated by maxima of the electric field which is resonantly enhanced within the cavity or by maximum absorption. An example is shown in Fig. 5.11, where the electron density perturbation (a), the field enhancement (b) and the absorption coefficient (c) are plotted versus the electron thermal velocity ($0.01 < \dot{v}_T/c < 0.1$). The caviton profile (d) used in our calculations is composed of a linear profile with $L = 3\lambda_0$ and a small cavity ($\delta N/N_0 < 0.05$) near the critical density. The coincidence of the peaks in all of the three graphs clearly indicates the operation of electrostatic resonances. This is further supported by the fact that the absorption at the main resonance exceeds the optimum value of 50 % for linear profiles (see also /14/).

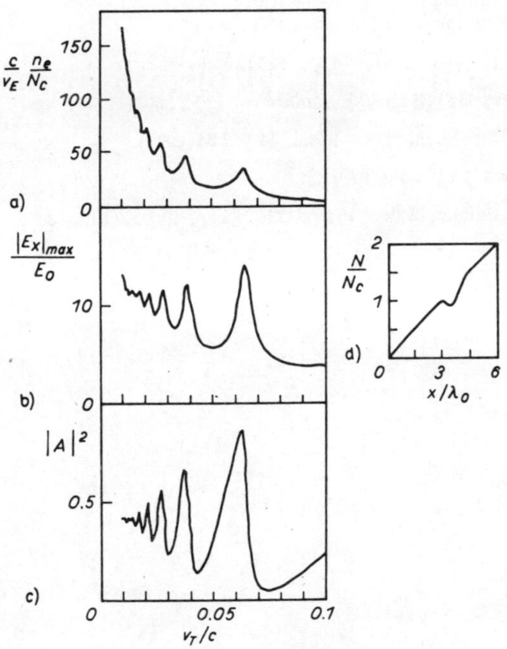

Fig. 5.11 Illustration of electrostatic structure resonances: electron density perturbation (a), longitudinal field enhancement (b) and absorption coefficient (c) versus the normalized electron thermal velocity v_T/c for the caviton density profile shown in (d); $\theta = 15^o$

Using a similar model, the consequences of electrostatic structure resonances on the effectivity of second harmonic generation in inhomogeneous plasmas are discussed in the next chapter.

References

/1/ Brekhovskikh, L.M.; Waves in Layered Media, Academic Press, New York, 1960.

/2/ Knittl, Z.; Optics of Thin Films, London, 1976.

/3/ Aliev, Yu.M., Vukovich, S., Gradov, O.M., Kyrij, A.Yu., Cadez, V.M.; Phys. Rev. A 15, 2120(1977).
Pisma Zh. Eksp. Teor. Fiz. 25, 351(1977).

/4/ Sauer, K.; Phys. Lett. 66A, 37(1978).

/5/ Sauer, K., Andreev, N.E., Baumgärtel, K.; Proc. VII. Conf. Plasma Phys. Contr. Nucl. Fusion Res., Innsbruck 1979, p. 187.

/6/ Mayer, F.J., Osborn, R.K., Daniels, D.W., McGrath, J.F.; Phys. Rev. Lett. 40, 30(1978).

/7/ Charov, A.A., Kondratev, I.G., Miller, M.A.; Pisma Zh. Eksp. Teor. Fiz. 25, 355(1977).

/8/ Sauer, K., Gorbunov, L.M.; Fiz. Plazmy 3, 1302(1977).

/9/ Andreev, N.E., Sauer, K.; Beitr. Plasmaphys. 17, 293(1977).

/10/ Sauer, K., Baumgärtel, K., Auer, G.; Phys. Lett. 74A, 211(1979).

/11/ Andreev, N.E., Gradov, O.M., Sünder, D.; Beitr. Plasmaphys. 20, 201(1980).

/12/ Kindel, J.M., Lee, K., Lindman, E.L.; Phys. Rev. Lett. 34, 134(1977).

/13/ Maki, H., Niu, K.; J. Phys. Soc. Japan 45, 269(1978).

/14/ Broskij, Yu.Ya., Goltsman, V.L., Netchuev, S.I.; Pisma Zh. Eksp. Teor. Fiz. 24, 547(1976).

Part III High-frequency nonlinear processes

6 Harmonic generation

When an intense electromagnetic wave of frequency ω impinges on an overcritical plasma, higher harmonics $n\omega$, $n = 2, 3, \ldots$ can be observed in the spectrum of the reflected radiation. Harmonic generation was investigated in a number of experiments on laser-produced plasmas /1 - 15/ and at the interaction of strong microwaves with plasmas /16, 17/. Among them, there are time and space resolved measurements /10 - 12/ and the observation of backscattered harmonics up to 5ω /13/ and 11ω /12, 14/, respectively, at laser intensities with $I\lambda^2$ of the order of 10^{16} W $\mu m^2/cm^2$. Recently published results describe observations up to the 46th harmonic at high power CO_2 laser facilities /18, 19/. Since the harmonic radiation is created mainly near the critical density, it is of importance for the diagnostics of processes taking place at the critical sheath /20 - 25/. So, it can help us understand the mechanisms involved in the absorption of laser radiation.

The process of harmonic generation in plasmas for not too high amplitudes of the incident field is analogous to the corresponding process in nonlinear optics /26/. The fundamental mechanism can easily be understood in terms of the generalized wave Eq. (2.12). An incident pump wave of frequency ω creates sources with 2ω, 3ω, ... via nonlinear terms of second, third, ... order, respectively. The excited modes may interact again with the pump to produce further combination frequencies. When appropriately polarized, the nonlinear sources generate transverse waves that can leave the plasma. Representing the total electric field as sum of the fields associated with the fundamental wave and the harmonics, a set of wave equations for the frequencies $n\omega$, $n = 1, 2, \ldots$ results from Eq. (2.12), coupled through nonlinear source terms containing products of fields varying with frequencies $m\omega$, $m \neq n$.

For efficient harmonic generation to occur, however, the existence of corresponding nonlinear source terms is not sufficient. Matching conditions have to be fulfilled in addition which are related to the spatial variation of the waves involved in the interaction. The wave number $k(n\omega)$ associated with the generated frequency $n\omega$ through the dispersion relation is required to coincide with the wave number describing the spatial variation of the source term. With other words, both frequency- and wave number matching conditions between the interacting waves must be satisfied in order to guarantee efficient harmonic generation.

A simple electromagnetic wave propagating in a homogeneous plasma cannot generate its second harmonic because of a lack of nonlinear coupling between the two transverse waves. If, however, a longitudinal plasma wave at the pump frequency is excited by the transverse electromagnetic wave, then the interaction of this electrostatic wave with either the pump or itself may generate a transverse harmonic. An inhomogeneous plasma provides a situation favourable for harmonic generation, since both, a longitudinal wave at the pump frequency can be generated by linear mode conversion and the wave number matching relation can be satisfied locally because of the different dispersion of longitudinal and transverse waves.

In principle, there are two mechanisms that generate longitudinal waves with a frequency near to the frequency of the incident wave: (i) resonance absorption as described in Section 4.3, which involves the linear mode conversion of a obliquely incident p-polarized wave at the critical density N_c and (ii) the parametric decay, in which the incident electromagnetic wave decays into an electron plasma wave of nearly the same frequency and a low-frequency ion acoustic wave /27/. In the latter case the second harmonic radiation is red-shifted by the ion sound frequency and because of its dependence on the electron temperature, the observation of the shift may be useful for diagnostic purposes /24, 25/. The second harmonic (SH) radiation in this case, however, will occur only when the amplitude of the pump exceeds the threshold for parametric decay, which causes this mechanism to become effective at higher intensities only.

Higher harmonic generation based on mode conversion in inhomogeneous plasmas were first investigated in /28 - 30/. Since the dispersion and dissipation of electron plasma waves depend very much on the electron temperature, the generation of harmonics should also be heavily affected. First analytical studies on this problem were carried out by Vinogradov and Pustovalov /28/. Later on, Erokhin et al. /30/ showed that the intensity of the SH emission first increases strongly with temperature, but oscillates at higher temperatures. The analytical calculations they made resulted in various asymptotic expressions that are valid in different regimes. The temperature dependence of the SH intensity over the full range of parameters involved, i. e., density scale length, angle of incidence and temperature, is not easily ascertained from these formulas. A theory for the emission of higher harmonics than the second in a cold plasma was worked out by Vladimirsky and Silin /31/.

In this chapter we study the temperature dependence of both, the second harmonic as well as the third in plasmas with linear density profiles. This is achieved by numerical solution of the wave equation for the fundamental wave including thermal effects, and the wave equations for the second and third harmonics with nonlinear source terms. The effect of Landau damping is included, leading to a strong decrease of the harmonic emission at higher temperatures.

Another purpose of this study is to point out the importance of density profile modifications on harmonic generation /30, 32/. Such modifications may result from the action of the ponderomotive force of the incident p-polarized wave. In a density cavity near the critical sheath electrostatic structure resonance as discussed in Section 5.3 with highly localized electrostatic fields can build up and lead to a significant increase in harmonic emission. For a rigid non-monotonous density profile these resonances can be found by varying the electron temperature. The formation of a density cavity, however, is a nonstationary process with a time scale determined by the ion motion (cf. Chapter 11) and harmonic radiation may indicate transient resonance configurations by enhanced emission. We suggest that this mechanism may be responsible for the peaks in time-resolved measurements for the second harmonic /10, 33/. Enhanced harmonic emission due to resonant excitation of surface waves in a cold plasma (cf. Section 5.2) was studied in /34/.

In Section 6.4 an alternative mechanism for harmonic generation operating at high intensities of the incident radiation will briefly be mentioned that is based on the anharmonicity of the electron motion near the critical density under the influence of the incident wave.

6.1 Calculation of the nonlinear source terms

We consider the oblique incidence of a p-polarized electromagnetic wave upon a thermal plasma. Since the pump wave interacts either with itself or with a longitudinal wave generated by mode conversion, harmonic generation does not need a noise level of any disturbance in the plasma to be initiated and thus no threshold exists for this process. We may assume stationary conditions and represent the total electric field as superposition of the fundamental wave (n = 1), the second harmonic (n = 2) and the third harmonic (n = 3)

$$\vec{E}(\vec{r}, t) = \mathrm{Re}\left\{\sum_{n=1}^{3} \vec{E}_n(\vec{r}) e^{-in\omega t}\right\} \tag{6.1}$$

No assumption is made concerning the spatial dependence of the complex amplitudes $\vec{E}_n(\vec{r})$. Similarly we expand $\vec{v}_e$ as

$$\vec{v}_e(\vec{r}, t) = \mathrm{Re}\left\{\sum_{n=1}^{3} \vec{v}_{en}(\vec{r})\, e^{-in\omega t}\right\}. \tag{6.2}$$

Inserting both expressions in the generalized wave equation (2.12) and arranging the terms according to their time-dependence yields the three wave equations /36/

$$-c^2 \mathrm{rotrot}\, \vec{E}_n + 3v_T^2 \mathrm{graddiv}\, \vec{E}_n + n^2\omega^2 \varepsilon(n\omega, N)\, \vec{E}_n = \vec{Q}_n , \tag{6.3}$$

n = 1, 2, 3, where $\vec{Q}_n$ denotes the nonlinear source terms.

The further analysis is simplified by the following assumptions: (a) Thermal effects can be neglected in the wave equations for the harmonics because they are generated near the critical layer associated with the pump frequency, $N \lesssim N_c$ and thus far from their own regions of resonance ($N \lesssim 4N_c$, $N \lesssim 9N_c$). (b) We neglect any reaction of higher harmonics on lower ones, i. e., of the second and third on the fundamental and of the third on the second. This is equivalent to a description of the process in terms of a higher-order perturbation theory, valid for not too high amplitudes of the incident wave. Quantitatively, the intensities $I_n \propto |\vec{E}_n|^2$ of the harmonics have to satisfy the inequality

$$I_1 \gg I_2 \gg I_3 .$$

Under these assumptions the source term in the wave equation for the fundamental wave can be dropped and we obtain for $\vec{E}_1$ the equation

$$- c^2 \operatorname{rotrot} \vec{E}_1 + 3v_T^2 \operatorname{graddiv} \vec{E}_1 + \omega^2 \varepsilon(\omega, N)\, \vec{E}_1 = 0 \tag{6.4}$$

which is identical to Eq. (4.2). Neglecting thermal effects leads to the equations

$$- c^2 \operatorname{rotrot} \vec{E}_n + n^2\omega^2 \varepsilon(n\omega, N)\, \vec{E}_n = \vec{Q}_n , \; n = 2, 3, \tag{6.5}$$

for the second and third harmonic.

In order to calculate the source terms $\vec{Q}_2$, $\vec{Q}_3$ we substitute the expansions (6.1), (6.2) into the expression (2.13) and find

$$\vec{Q}_2 = \frac{m_e}{e}\omega_p^2 \frac{1}{4} \operatorname{grad} (\vec{v}_{e1} \vec{v}_{e1}) - i\omega\vec{v}_{e1} \operatorname{div} \vec{E}_1 , \tag{6.6}$$

$$\vec{Q}_3 = \frac{m_e}{e}\omega_p^2 \frac{1}{2} \operatorname{grad} (\vec{v}_{e1} \vec{v}_{e2}) - \frac{3i\omega}{2} (\vec{v}_{e2} \operatorname{div} \vec{E}_1 + \vec{v}_{e1} \operatorname{div} \vec{E}_2) . \tag{6.7}$$

$\vec{v}_{e1}$, $\vec{v}_{e2}$ are eliminated with the help of the equation of motion for the electrons in the version (2.1), that is,

$$- i\omega\vec{v}_{e1} = \frac{e}{m_e} \vec{E}_1 , \tag{6.8}$$

$$- 2i\omega\vec{v}_{e2} = \frac{e}{m_e} \vec{E}_2 - \frac{1}{4} \operatorname{grad} (\vec{v}_{e1} \vec{v}_{e1}) . \tag{6.9}$$

This yields the final form /36/

$$\vec{Q}_2 = \frac{e}{m_e}\left[\vec{E}_1 \operatorname{div} \vec{E}_1 + \frac{1}{4}\frac{N}{N_c} \operatorname{grad} (\vec{E}_1 \vec{E}_1)\right] , \tag{6.10}$$

$$\begin{aligned} \vec{Q}_3 = - \frac{e}{m_e} \Bigg[& \frac{1}{4}\frac{N}{N_c} \operatorname{grad} (\vec{E}_1 \vec{E}_2) + \frac{3}{4}\vec{E}_2 \operatorname{div} \vec{E}_1 \\ & - \frac{e}{16\, m_e \omega^2}\frac{N}{N_c} \operatorname{grad} \left(\vec{E}_1 \operatorname{grad} (\vec{E}_1 \vec{E}_1)\right) \\ & - \frac{3\, e}{16\, m_e \omega^2} \operatorname{grad} (\vec{E}_1 \vec{E}_1) \operatorname{div} \vec{E}_1 + \frac{3}{2} \vec{E}_1 \operatorname{div} \vec{E}_2 \Bigg] \end{aligned} \tag{6.11}$$

of the source terms for the second and the third harmonic. We use the coordinate system introduced in Section 4 with the x-direction along the density gradient and with the x-z-plane being the plane of incidence. The electric fields can then be written

$$\vec{E}_n(\vec{r}) = \vec{E}_n(x, z) = \left[E_{nx}(x)\,\vec{e}_x + E_{nz}(x)\,\vec{e}_z \right] e^{ik_{nz}z} \tag{6.12}$$

with

$$k_{1z} = k_o \sin\theta ,$$

where k_o is the vacuum pump wave number and θ is the angle of incidence. Because of $\vec{E}_1 \propto \exp(ik_1 z)$, the source terms $\vec{Q}_2$, $\vec{Q}_3$ exhibit the z-dependencies

$$\vec{Q}_2 \propto e^{2ik_1 z} , \quad \vec{Q}_3 \propto e^{3ik_1 z} ,$$

and thus the wavenumber resonance condition in z-direction requires

$$k_{2z} = 2k_{1z} , \quad k_{3z} = 3k_{1z} .$$

The wave equations (6.4), (6.5) represent a system of ordinary differential equations of second order for $E_{nx}(x)$, $E_{nz}(x)$, $n = 1, 2, 3$ which can be solved by standard procedures (discretization and vector Gauss elimination /35/). The boundary conditions follow from the continuity conditions for the electric and magnetic fields. The inclusion of Landau damping in the equation for the fundamental wave via the phenomenological coefficient ν_L (cf. Eq. (4.15)) guarantees that the generated plasma waves are damped out before they can reach the boundary. The theory is controlled by three parameters: The angle of incidence θ, the ratio v_T/c as measure of the electron temperature and the ratio L/λ_o of the density gradient scale length to the free space wavelength of the pump. Numerical results are reviewed in /36/.

6.2 Linear density profile

In Fig. 1 typical field patterns of both the fundamental wave and the second harmonic (SH) are shown in the case of a linear density profile

$$N(x) = \frac{x}{L} N_c$$

for $L = 4\lambda_o$, $v_T/c = 0.05$. The angle of incidence, $\theta = 20^o$, has been chosen such that approximately maximum resonance absorption occurs. The spatial amplitude profiles of the fundamental wave (Figs. 1 (a) - (c)) are identical to those shown in Fig. 4.6 and we refer to the comments given there. In Figs. 1 (d), (e) the associated SH field components are plotted. As may be seen, the second harmonic is emitted in direction of lower density only. This is a typical feature of a linear density profile and is in agreement with analytical results /30, 31/. The correlation between Figs. 1 (a) and (b) and Figs. 1 (d) and (e), respectively, is obvious. It reflects the dominant role of the nonlinear term $\vec{E}_1 \operatorname{div} \vec{E}_1 \propto \vec{E}_1 n_{e1}$ for the second harmonic generation at a linear density profile.

For the presentation of further results concerning the efficiency of harmonic generation we introduce the dimensionless quantities

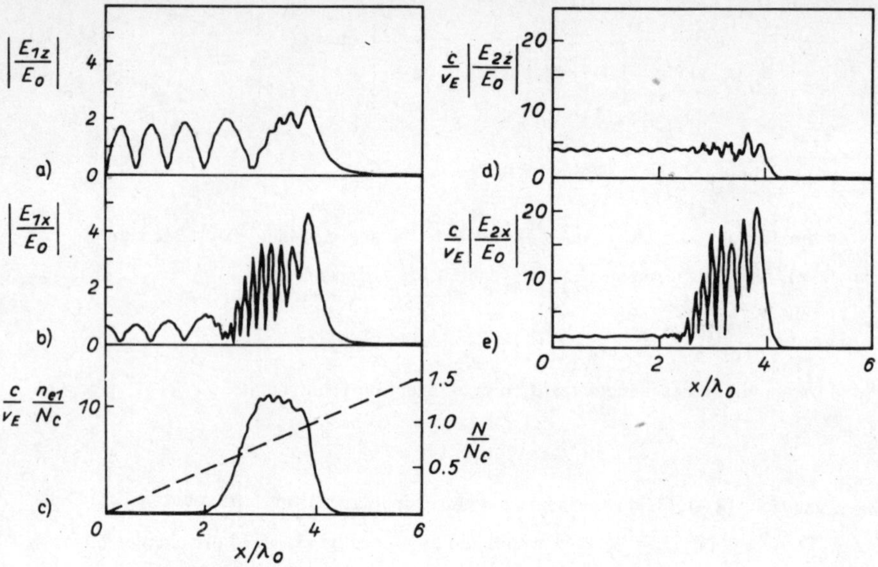

Fig. 6.1 Spatial amplitude profiles of the electric field of the incident wave, associated electron density perturbation and electric field of the second harmonic (SH) for a linear plasma density profile with $L = 4\,\lambda_o$, $\theta = 20^o$; electric field component transverse (a) and parallel (b) to the density gradient, electron density perturbation (c) and undisturbed density profile (dashed line), SH field component transverse (d) and parallel (e) to the density gradient.
The first order fields are normalized by E_o, the SH fields by $(v_E/c)\,E_o = eE_o^2/m_e\omega c$ and the density perturbation by $(v_E/c)\,N_c = eE_oN_c/m_e\omega c$.

$$u_n = \frac{e|\vec{E}_n|}{m\omega c} . \tag{6.13}$$

Since in the case of negligible re-action of the higher harmonis the vacuum intensity $I_n = |u_n(0)|^2$ of the n-th harmonic is proportional to the n-th power of the vacuum pump intensity $I_o = |u_o(0)|^2$, i. e., $I_n \propto I_o^n$, we will present our results in terms of the ratio

$$K_n = I_n/I_o^n, \quad n = 2, 3 \tag{6.14}$$

because this quantity will be independent of the intensity of the incident wave. The coefficient of energy transfer to the harmonics (or the conversion factor used by Basov et al. /15/), i. e., the intensity of the emitted harmonics relative to the incident intensity

$$Q_n = I_n/I_o \tag{6.15}$$

is then obtained from K_n by forming the product

$$K_n I_o^{n-1} = Q_n . \tag{6.16}$$

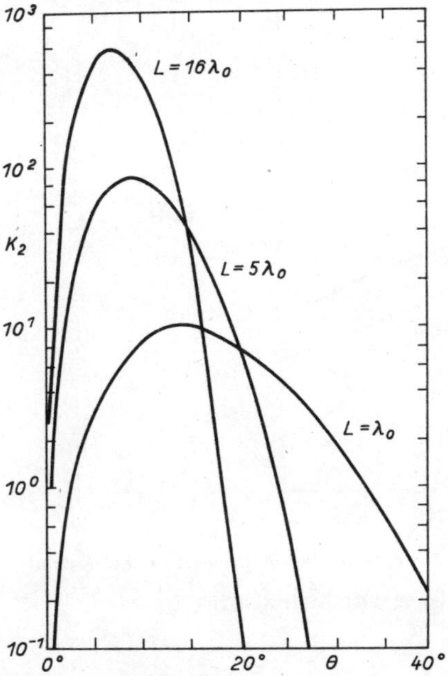

Fig. 6.2 SH transformation coefficient K_2 versus the angle of incidence for three density scale lengths L, $v_T/c = 0.05$

In Fig. 6.2 the coefficient K_2 for the second harmonic is plotted as function of the angle of incidence for three density scale lengths (the electron temperature is kept constant, $v_T/c = 0.05$). Like the angle for maximum resonance absorption (cf. Fig. 4.7), the angle for maximum second second harmonic emission is also shifted toward smaller values when the scale length increases. In particular, we should point to the following fact: For $L \approx 20\,\lambda_o$, a value of K_2 of the order of 10^3 is found. This means, for example, that for $u_o = 0.01$ a second harmonic of 10 % of the incident intensity is achieved. Therefore, we can expect that for such gradients, saturation effects resulting from the re-action of the second harmonic back on the pump will already occur at amplitudes of $u_o \gtrsim 0.01$.
An interesting feature of the SH generation is shown in Fig. 6.3 where the coefficient K_2 is plotted as function of the electron temperature for three density scale lengths and for the optimum angle which is taken from Fig. 6.2. The plots clearly show the strong influence of the electron temperature especially in the case of weak gradients ($k_oL \gg 1$). Starting from the cold plasma values /29, 34/, the coefficient K_2 increases first with temperature, i. e., with v_T/c. Having passed through a maximum, an oscillating behaviour sets in which for $v_T/c > 0.15$ goes over into an exponential decay. The oscillations are the more distinct the longer the density scale length becomes.

As was first pointed out by Erokhin et al. /29/, these oscillations are a direct consequence of the fact that in an inhomogeneous plasma the wave number matching condition for the dominant process $T + L \longrightarrow T$,

$$k_x^{(T)}(2\omega, x) = k_x^{(T)}(\omega, x) + k_x^{(L)}(\omega, x, v_T/c), \tag{6.17}$$

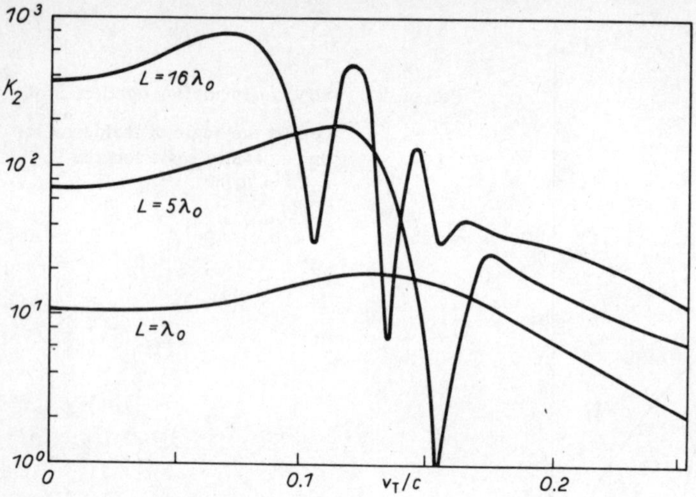

Fig. 6.3 SH transformation coefficient K_2 at optimum angle of incidence versus the normalized electron thermal velocity v_T/c for three density scale lengths L

is fulfilled only locally, i. e., at a certain point $x_R^{(2)}$. By using the dispersion relation for longitudinal and transverse waves (Eqs. (4.6) and (4.7)), its distance from the critical sheath is found to be

$$d^{(2)} = x_c - x_R^{(2)} \simeq 9\,(v_T/c)^2\,L\,. \tag{6.18}$$

With increasing temperature the matching point $x_R^{(2)}$ moves toward lower densities, thereby passing the standing wave structure of the fundamental transverse wave which is not enfluenced by the electron temperature (Fig. 6.1 (a)). When it coincides with a node of $|E_{1z}|$, the coefficient K_2 will assume a minimum value, whereas in the opposite case it will pass a maximum. In this way, high and low values of K_2 alternate until the longitudinal wave decreases because of Landau damping. Thus, the temperature dependence of K_2 is correlated with the spatial structure of the product $n_{el}\ E_{1z}$ which plays the decisive role in the SH generation.

According to this picture, we also expect oscillations of K_2 when the density scale length is changed at a fixed temperature. This effect is shown in Fig. 6.4, where K_2 is plotted as a function of L for two values of the temperature (v_T/c = 0.05 and 0.15, respectively). Especially for the higher temperature strong oscillations of K_2 are found when L is varied.

For the third harmonic, similar effects occur. It is no longer possible, however, to correlate the temperature dependence of K_3 (see Fig. 6.5) unambiguously with a single term of the source $\vec{Q}_3$ which is given by Eq. (6.11). For large scale lengths ($L > 10\lambda_o$) we observe oscillations of K_3 already at low temperatures ($v_T/c \approx 0.05$). The periodicity of these oscillations is rather short. They are apparently connected to the spatial pattern of the longitudinal component E_{1x}. At higher temperatures K_3 reproduces the structure of the transverse component E_{1z} in analogy to the behaviour of the second harmonic. In this temperature region the third harmonic generation seems to be determined by the process T + L + L ⟶ T accord-

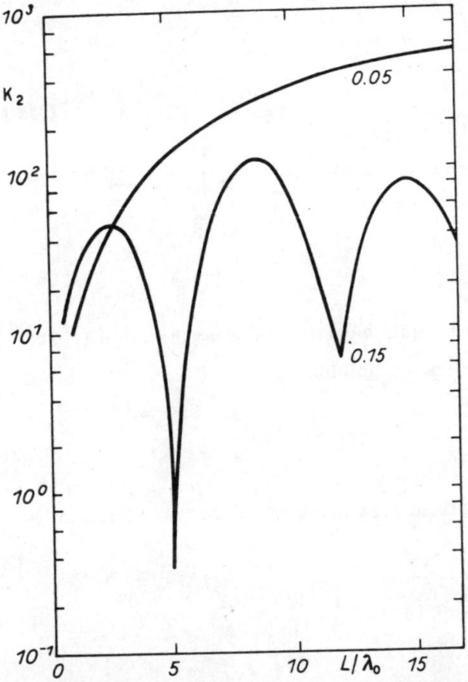

Fig. 6.4 SH transformation coefficient K_2 versus the density scale length L for two values of the electron temperature (v_T/c)

ing to the source term $n_{el} n_{el} E_{1z}$. Then, the distance between the critical point and the matching point $x_R^{(3)}$ for this process can be calculated once more from the wave number matching relation

$$k_T(3\omega, x) = 2 k_L(\omega, x, v_T/c) + k_T(\omega, x) . \tag{6.19}$$

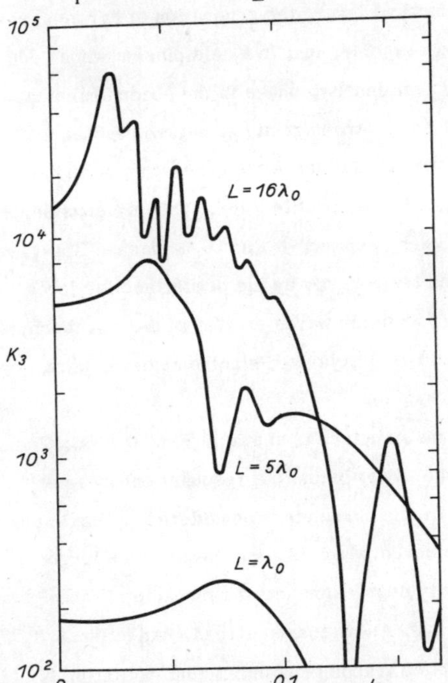

Fig. 6.5 Third harmonic transformation coefficient K_3 at optimum angle of incidence versus the electron thermal velocity for three density scale lengths L

In analogy to Eq. (6.18) one obtains

$$d^{(3)} = x_c - x_R^{(3)} \simeq 6\,(v_T/c)^2\,L\,. \tag{6.20}$$

If the point $x_R^{(3)}$ coincides with a node of the standing wave pattern of $|E_{1z}|$, K_3 will become a minimum. The same node, however, may also cause a minimum value of K_2, if

$$x_R^{(2)} = x_R^{(3)} \tag{6.21}$$

is fulfilled. When L is given, from Eq. (6.21) we obtain a relation between the value $(v_T/c)^{(2)}$, for which K_2 is minimal, and $(v_T/c)^{(3)}$, for which K_3 is minimal

$$(v_T/c)^{(3)} \simeq \sqrt{1.5}\;(v_T/c)^{(3)}\,. \tag{6.22}$$

Considering the numerical results for $L = 16\,\lambda_o$, from Figs 6.3 and 6.5 we find minima of K_2 and K_3 at

$$(v_T/c)^{(2)} \simeq 0.105\,,\ 0.135 \text{ and } (v_T/c)^{(3)} \simeq 0.13\,,\ 0.165$$

which indeed fulfil the relation (6.22).

6.3 Non-monotonous density profiles

When a linear density profile is deformed near the critical layer, the generation of harmonics may be affected very significantly /30/. It was shown in /34/, that in a cold plasma whose density profile exhibits a plateau structure near the critical density, peaks in the second harmonic emission may be created which are directly related to electromagnetic structure resonances. Similarly, electrostatic structure resonances may strongly influence the generation of higher harmonics in warm plasmas having a non-monotonous density profile /36/. When we consider a fixed density profile with a cavity near the critical layer as shown in Fig. 6.6 (dashed line), we want to simulate density structures which are formed temporarily by the ponderomotive force associated with the high-frequency short wavelength Langmuir waves excited in the neighborhood of the resonance layer in the case of an obliquely incident p-polarized electromagnetic wave (cf. Chapter 11).

A typical example of such an electrostatic resonant structure is shown in Fig. 6.6 (a). The superposition of the electrostatic waves trapped in the cavity leads to a resonant enhancement of the longitudinal component of the electric field. For the parameters considered, a maximum enhancement $|E_{1x}|/E_o$ of approximately 12 is achieved which is about three times larger than that for a linear profile with $L = \lambda_o$. The density depression acts a resonantor for the electrostatic waves which are excited by the pump wave. As discussed already in Section 5.3, this effect is optimized when a balance between complete trapping and maximum excitation is

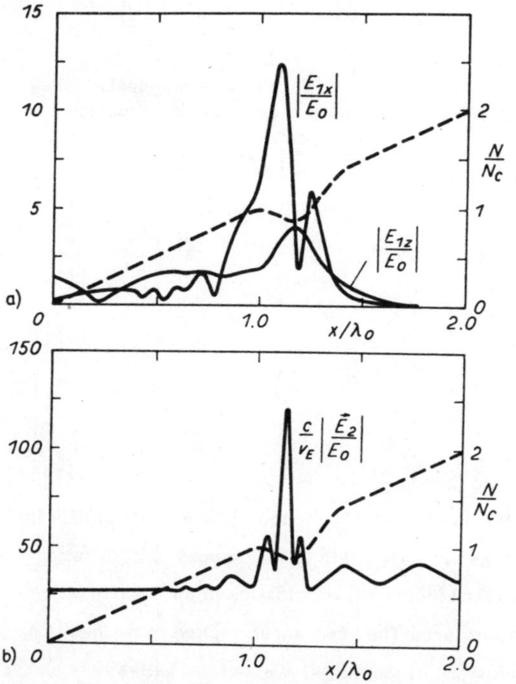

Fig. 6.6 Spatial amplitude profiles of the electric fields for a resonant caviton density profile ($v_T/c = 0.05$, $\theta = 20^o$); (a) transverse (E_{1z}) and parallel (E_{1x}) electric field components of the incident wave, (b) electric field amplitude of the second harmonic. The dashed line represents the undisturbed density profile, normalization as in Fig. 6.1

established. Therefore, strong resonances occur for just such density profiles where the density on the low-density side of the cavity is close to the critical value. (Similar effects were already observed in /37/.)

The influence of resonant electrostatic structures on the SH generation can be seen in Fig. 6.6 (b), where the amplitude of the second harmonics is plotted. The presence of trapped longitudinal waves not only leads to an increase of the SH emission, but also results in an almost symmetric emission of the SH radiation in both directions, contrary to a linear profile, where the second harmonic is emitted only in the backward direction.
It should be possible to use this effect for an indirect demonstration of self-formed caviton structures when laser radiation interacts with a plasma that is transparent to the second harmonic. In time-resolved measurements, SH emission in the forward direction should occur at those moments only when a resonant cavity is formed. In this way, useful information on plasma dynamic processes in the vicinity of the critical sheath could be obtained.

Nonstationary self-consistent calculations of the interaction of a p-polarized wave with an inhomogeneous plasma show, that enhanced absorption and enhanced harmonic emission are correlated /38/. This is easily understood as a consequence of the temporary formation of resonant cavities.

The intrinsic electrostatic character of the resonances under discussion is manifested once more in Fig. 6.7 which shows the spatial structure of the electron density perturbation $n_{e1} \propto \mathrm{div}\, \vec{E}_1$. We can see that the oscillations resulting from the trapping of the electrostatic waves within the density cavity are directly reproduced by the spatial structure of the SH amplitude of Fig. 6.6 (b). The maximum density perturbation n_{e1} is three times larger than in the

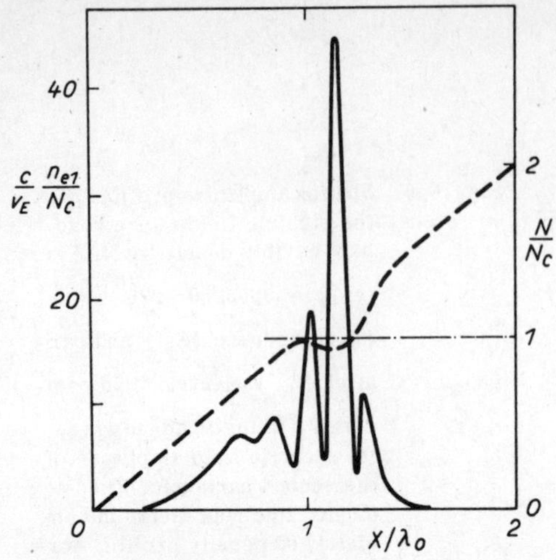

Fig. 6.7 Spatial profile of the electron density perturbation for a caviton density profile (dashed line), parameters as in Fig. 6.1

case of a strictly linear profile of the same scale length.

The effect of the electron temperature on the coefficient K_2 for the same caviton profile is plotted in Fig. 6.8. With increasing temperature the wavelength of the trapped electrostatic waves increases too. Pronounced peaks of K_2 then indicate the formation of an electrostatic structure resonance at the corresponding temperature. The resonances related to the maxima of K_2 at $v_T/c \simeq 0.045$ and 0.02 are characterized by one half-wave and two half-waves just fitting into the density cavity. The secondary maximum at $v_T/c \simeq 0.065$ apparently corresponds to the side maximum of n_{e1} (see Fig. 6.7). At higher temperatures the wavelengths of the electrostatic waves become larger than the dimension of the density depression as a consequence of which the caviton loses its influence on the field enhancement and second harmonic

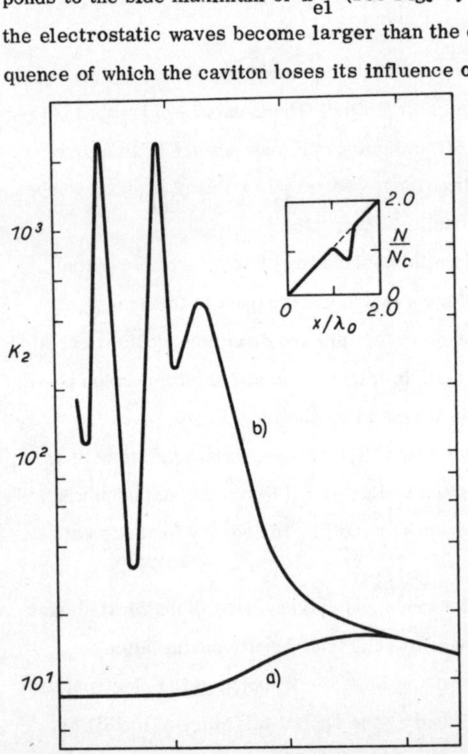

Fig. 6.8 Enhanced SH generation by electrostatic caviton resonances; SH transformation coefficient K_2 versus the electron thermal velocity for two plasma density profiles, (a) linear profile with $L = \lambda_o$, (b) caviton profile as shown in Fig. 6.6 and 6.7. The peaks are associated with caviton resonances

generation. Therefore, with increasing temperature K_2 gradually approximates the value of K_2 of a plasma with a strictly linear profile

6.4 Harmonic generation by anharmonic electron motion

In recent experiments on CO_2-laser-produced plasmas, observations of up to the 46th harmonic have been reported /18, 19/. The results show that the harmonic production efficiency is constant over the harmonic number with a sharp cutoff at the highest harmonic. This behaviour is in contrast with the predictions of the higher-order perturbation theory developed above. A theoretical explanation of this highly nonlinear nature of harmonic generation has been given in terms of the anharmonicity of the electron motion near the critical density at a very sharp density gradient /39 - 42/ Fig. 6.9 illustrates the anharmonicity that gives rise to harmonic generation. Electrons belonging to the critical layer oscillate under the influence of the longitudinal field generated by the incident wave via mode conversion (p-polarization) up and down the density gradient with the local plasma frequency. Such an electron, attempting to oscillate within the thin region of the sharp density gradient sees a highly asymmetric environment. It moves from low to high density, and, since the electrons of a plasma wave see a restoring force that is essentially proportional to the plasma frequency, from a region of low restoring force to one of

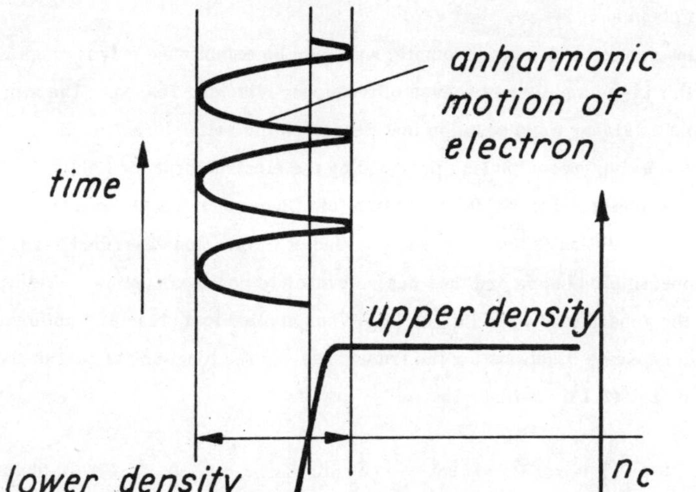

Fig 6.9 Schematic representation of the anharmonic motion of an electron at critical density within the thin region of a sharp density gradient, leading to higher harmonie generation

high restoring force. Thus, it experiences an extremely anharmonic acceleration. The result of this anharmonicity is the generation of scattered light containing the higher harmonics of the incident radiation. Thereby, the highest possible harmonic frequency is limited by the highest plasma frequency associated with the density profile, i. e., the plasma frequency of the upper density shelf. Because the density of the plasma is proportional to ω_p^2, the observation of the 40th harmonic implies a density of the upper shelf that is 1600 times N_c.

Sharp density gradients may be created self-consistently by the ponderomotive force of the incident wave field (cf. Chapters 10,11). For example, Fig. 10.11 illustrates a situation where a density jump with a characteristic scale length of about one vacuum wave length separates a lower and an upper density shelf.

6.5 Self-interaction of transverse waves in a homogeneous plasma

Harmonic generation is one of the high-frequency nonlinear processes that occurs when an electromagnetic wave is incident upon a nonuniform plasma. The phenomenon was analyzed in the previous sections within the framework of a higher-order perturbation theory. The reaction of the nonlinearly generated modes back on the fundamental wave has been ignored in the analysis. In this section we pay attention particularly to this effect, looking at harmonic generation from the point of view of a self-interaction of an almost monochromatic transverse wave propagating in a homogeneous plasma.

It is obvious that self-interaction of a monochromatic wave can be established only through a nonlinear term of third order in the amplitude because of frequency matching reasons. The aim of this section is to derive a nonlinear wave equation that describes the self-interaction of a transverse electric field in a homogeneous plasma provided by the electron dynamics only. It has some similarity with the nonlinear Schrödinger equation (cf. Chapter 8), but there is a distinctive difference. Since ion motion is not involved, the kind of interaction described here can occur also in an incompressible plasma and thus has relevance to nonlinear optics.

Starting point is again the generalized wave Eq. (2.12). When applied to a plasma of uniform density it splits up into two separate equations for the transverse and the longitudinal part of the electric field. Taking rot of Eq. (2.12) we find

$$\left(-\frac{\partial^2}{\partial t^2} + c^2\triangle - \omega_p^2 \right) \quad \mathrm{rot}\,\vec{E} = \mathrm{rot}\,\frac{\partial}{\partial t}(\vec{v}_e\,\mathrm{div}\,\vec{E})\,. \tag{6.23}$$

This equation governs the propagation and generation of transverse disturbances in a plasma of uniform density. From Eq. (6.23) it is obvious that in a homogeneous plasma sources for electromagnetic disturbances can only be created when a longitudinal electric field is present. This means particularly that a monochromatic transverse wave is unable to produce its second harmonic when propagating in a uniform plasma, as pointed out already in Section 6.1. Another consequence from Eq. (6.23) is that self-interaction of a transverse wave can only operate via a longitudinal field of either twice the fundamental frequency or zero frequency.

As far as the generation of a longitudinal wave with 2ω is concerned, it can easily be seen that wave number matching cannot be satisfied. The source term varies in space proportional $\exp(2ik_T(\omega)x)$ where $k_T = (\omega^2 - \omega_p^2)^{1/2}/c$ is the transverse wave number. The wave number belonging to a Langmuir wave of frequency 2ω, however, is $k_L(2\omega) = ((2\omega)^2 - \omega_p^2)^{1/2}/\sqrt{3}\, v_T$ and we have $k_L(2\omega) \gg 2k_T(\omega)$ due to $c \gg v_T$. Hence, longitudinal waves of twice the fundamental frequency cannot be excited resonantly. The key for the self-interaction mechanism is the excitation of a longitudinal zero-frequency disturbance $\vec{E}_o$ in the plasma by the transverse wave.

In Chapter 3 it was pointed out already that a quasistationary electric field may be produced by a high-frequency field. Applying Eq. (3.7) to a homogeneous plasma we find

$$\frac{e}{m_e}\vec{E}_o = -\frac{1}{4}\,\mathrm{grad}\,|\vec{v}_{e\omega}|^2 \tag{6.24}$$

where

$$\vec{v}_{e\omega} = \frac{e\,\vec{E}_\omega}{i m_e \omega}, \quad \vec{E}(\vec{r}, t) = \mathrm{Re}\left\{\vec{E}_\omega e^{-i\omega t}\right\}. \tag{6.25}$$

Inserting Eqs. (6.24), (6.25) into Eq. (6.23) with $\mathrm{div}\,\vec{E}_\omega = 0$ and turning to the modulation representation we arrive at the nonlinear wave equation

$$2i\omega\frac{\partial \vec{E}_\omega}{\partial t} + c^2\Delta\vec{E} - \omega^2\mathcal{E}\vec{E}_\omega = \frac{1}{4}\,\frac{e^2}{m_e^2\omega^2}\Delta|\vec{E}_\omega|^2\,\vec{E}_\omega \tag{6.26}$$

for the slowly varying complex amplitude $\vec{E}_\omega$. The characteristic difference to the nonlinear Schrödinger equation (cf. Eq. (8.28)) ist the appearance of a second-order spatial derivative at the square of the absolut value of the amplitude.

It should be emphasized, however, that the kind of self-interaction described by Eq. (6.26) will be modified and possibly dominated by low-frequency processes when the plasma has time enough to respond to spatial amplitude gradients by ion-density profile modifications. The system (3.14) - (3.16) refers to such situations and the nonlinear Schrödinger equation evolves as particular case from this system Hence, Eq. (6.26) is applicable to those cases where, roughly speaking, the high-frequency amplitude structure moves rapidly enough (compared with the ion acoustic velocity) relative to the plasma. Self-interaction is then provided by the nonlinearities in the electron motion without affecting the ion density. Some phenomena of interest are the self-modulation of electromagnetic pulses propagating through long plasma wave guids or self-pulsing and frequency modulation in plasma-mirror systems which have their analogs in nonlinear optics /43 - 45/.

References

/1/ Bobin, J.L., Decroisette, M., Meyer, B., Vitel, Y. ; Phys. Rev. Lett. 30 , 594(1973).

/2/ Goldman, L.M., Soures, J., Lubin, M.J. ; Phys. Rev. Lett. 29 , 1184(1973).

/3/ Eidman, K., Sigel, R. ; Phys. Rev. Lett. 34 , 799(1975).

/4/ Krokhin, O.N., Pustovalov, V.V., Rupasov, A.A., Silin, V.P., Sklizkov, G.V., Starodub, A.N., Tikhonchuk, V.P., Shikanov, A.S. ; Pisma Zh. Eksp. Teor. Fiz. 22 , 47(1975).

/5/ Basov, N.G., Krokhin, O.N., Pustovalov, V.V., Rupasov, A.A., Silin, V.P., Sklizkov, G.V., Tikhonchuk, V.T., Shikanov, A.S. ; Zh. Eksp. Teor. Fiz. 67 , 118(1974).

/6/ Baldis, H.A., Pepin, H., Johnston, T.W., Parbhakar; Phys. Rev. Lett. 35 , 37(1975).

/7/ Aleksandrov, V.V., Vikharev, V.D., Zotov, V.P., Kovalski, N.G., Pergament, M.I. ; Pisma Zh. Eksp. Teor. Fiz. 24 , 551(1976).

/8/ Aleksandrov, V.V., Anisimov, S.I., Brenner, M.V., Velikhov, E.P., Vikharev, V.P., Zotov, V.P., Kovalski, N.G., Pergament, M.I., Yaroslavski, A.I. ; Zh. Eksp. Teor. Fiz. 71 , 1826(1976).

/9/ Baldis, H.A., Burnett, N.H., Enright, G.D., Richardson, M.C.; Appl. Phys. Lett. 34, 327(1979).

/10/ Carter, P.D., Sim, S.M.L., Hughes, T.P. ; Opt. Commun. 27 , 423(1978).

/11/ Leonhard, T.A., Cover, R.A. ; J. Appl. Phys. 50 , 3241(1979).

/12/ Jaanimagi, P.A., Enright, G.D., Richardson, M.C. ; IEEE Trans. Plasma Sci. PS-7 166(1979).

/13/ McLean, E.A., Stamper, J.A., Ripin, B.H., Griem, H.R., McMahon, J.M., Bodner, S.E. ; Appl. Phys. Lett. 31 , 825(1977).

/14/ Burnett, N.H., Baldis, H.A., Richardson, M.C., Enright, G.D. ; Appl. Phys. Lett. 31, 172(1977).

/15/ Basov, N.G., Bychenko, V.Yu., Krokhin, O.N., Osipov, M.V., Rupasov, A.A., Silin, V.P., Sklizkov, G.V., Starodub, A.N., Tikhonchuk, V.T., Shikanov, A.S. ; Zh. Eksp. Teor. Fiz. 76 , 2094(1979).

/16/ Batanov, G.M., Silin, V.A. ; Tr. Fiz. Inst. (Moskau) 73 , 87(1974).

/17/ Wong, A.Y. ; in "Laser Interaction and Related Phenomena", ed. by H.J. Schwarz, H. Hora; Plenum, New York, 1977, Vol. 4B, p. 783.

/18/ Carman, R.L., Forslund, D.W., Kindel, J.M.; Phys. Rev. Lett. 46 , 29(1981).

/19/ Carman, R.L., Rhodes, C.K., Benjamin, R.F. ; Phys. Rev. A 24 , 2649(1981).

/20/ Jackel, St., Albritton, J., Goldman, E. ; Phys. Rev. Lett. 35 , 514(1975).

/21/ Jackel, St., Perry, B., Lubin, M. ; Phys. Rev. Lett. 37 , 95(1976).

/22/ Saleres, A., Decroisette, M. ; Phys. Lett. A 51 , 379(1975).

/23/ Bychenkov, V.Yu., Zakhorenkov, Yu.A., Krokhin, O.N., Rupasov, A.A., Silin, V.P., Sklizkov, G.V., Starodub, A.N., Tikhonchuk, V.T., Shikanov, A.S. ; Pisma Zh. Eksp. Teor. Fiz. 26 , 500(1979).

/24/ Burges, M.D.J., Dragila, R., Luther-Davies, B. ; Optics Comm. 50 , 236(1984).

/25/ Tanaka, K., Seka, W., Goldmann, L.M., Richardson, M.C., Short, R.W., Sources, J.M., Williams, E.A. ; Phys. Fluids 27 , 2187(1984).

/26/ Blombergen, N. ; Nonlinear Optics, W.A. Benjamin, New York, 1965.

/27/ Silin, V.P. ; Parametric Effects of Intense Radiation on Plasmas (in Russ.), Nauka, Moscow, 1973.

/28/ Vinogradov, A.V., Pustovalov, V.V. ; Zh. Eksp. Teor. Fiz. 63 , 940(1972).

/29/ Erokhin, N.S., Zakharov, V E., Moiseev, S.S. ; Zh. Eksp. Teor. Fiz. 56 , 179(1969).

/30/ Erokhin, N.S., Moiseev, S.S., Mukhin, V.V. ; Nucl. Fusion 14 , 333(1974).

/31/ Vladimirski, A.B., Silin, V.P. ; Fiz. Plazmy 6 , 354(1980).

/32/ Baumgärtel, K., Sauer, K., Andreev, N.E., Auer, G. ; 13. Europ. Conf. on Laser Interaction with Matter Leipzig, 1979, paper K 38.

/33/ Gray, D.R., Murdoch, J., Sim, S.M.L., Cole, A.J., Evans, R.G., Toner, W.T. ; Plasma Phys. 22 , 967(1980).

/34/ Auer, G., Sauer, K., Baumgärtel, K. ; Phys. Rev. Lett., 42 , 1744(1979).

/35/ Richtmyer, R.D., Morton, K.W. ; Difference Methods for Initial Value Problems, Interscience, New York, 1967.

/36/ Andreev, N.E., Auer, G., Baumgärtel, K., Sauer, K. ; Phys. Fluids 24 , 1492(1981).

/37/ Morales, G.J., Lee, Y.C. ; Phys. Fluids 20 , 1135(1977).

/38/ Andreev, N.E., Silin, V.P., Stenchikov, G.L. ; Zh. Eksp. Teor. Fiz. 78 , 1396(1980).

/39/ Bezzerides, B., Jones, R.D., Forslund, D.W. ; Phys. Rev. Lett. 49 , 202(1982).

/40/ Grebogi, C., Tripati, V.K., Chen, H. ; Phys. Fluids 26 , 1904(1983).

/41/ Grebogi, C., Chen, H., Tripathi, V.K. ; in "Laser Interaction and Related Phenomena", ed. by H. Hora, G.H. Miley, Plenum, New York 1985, Vol. 6 , p. 401.

/42/ Forslund, D.W., Goldstone, Ph.D. ; Los Alamos Science No. 12, p. 2(1985).

/43/ DeMartine, F., Townes, C.H., Gustafson, T.K., Kelley, P.L. ; Phys. Rev. 164 , 312(1967).

/44/ Firth, W.J. ; Optics Commun. 39 , 343(1981).

/45/ Firth, W.J., Wright, E.M. ; Phys. Lett. 92 , 211(1982).

7 Stimulated Raman scattering (SRS) and two-plasmon decay (TPD)

When an intense electromagnetic wave is incident upon a nonuniform plasma, besides the harmonics $n\omega$ also the frequencies $\omega/2$, $3\omega/2$ may be observed in the backscattered radiation. These frequencies have been seen in a great number of experimental studies on laser-plasma interaction /1 - 25, 83 84/, including the investigation of their spectral, angular and temporal distribution. Among the mechanism that are responsible for the generation of these frequencies, stimulated Raman scattering and two-plasmon decay play a dominant role.

7.1 Parametric instabilities and three-wave coupling

Both processes are members of a family of parametric instabilities which can occur in an undercritical plasma when irradiated by an intense electromagnetic wave. Common to all of these processes is that the incident wave (the pump wave) drives two other waves through the coupling provided by the plasma. Thereby, frequency- and wavenumber matching relations

$$\omega = \omega_1 + \omega_2 , \tag{7.1}$$

$$\vec{k} = \vec{k}_1 + \vec{k}_2 \tag{7.2}$$

are satisfied between the three waves involved in the interaction. The basic mechanism behind this interaction is the strong nonlinear coupling that can occur when two waves beat together so that their sum or difference frequency and wavelenght just matches the frequency and the wavelength of a third wave. The process will be efficient if the pump wave is sufficiently strong to overcome the losses of the generated waves.

The excitation and amplification of plasma modes by an optical or microwave pumping may also be described in terms of a parametric interaction process. The pump wave modulates the plasma parameters which enter the dispersion relation of the natural plasma modes and instabilities resulting from this interaction have the character of parametric instabilities. The close connection between parametric instabilities in plasmas and resonant three-wave coupling results simply from the fact, that in several cases both terms describe basically the same process. A parametric instability may turn out to be a three-wave coupling process at its initial stage and consequently, three-wave coupling provides, among others, a possible saturation mechanism for a parametric instability.

A large number of resonant three-wave interactions in plasmas have been studied since the early 1960's (Oraevski and Sagdeev /26/, Silin /27/, Goldman /28/, DuBois and Goldman /29/). The general aspects of the theory of parametric excitation of coupled waves are surveyed by Sagdeev and Galeev /30/, Galeev and Sagdeev /31/ and Nishikawa /32/, who developed a general framework for the parametric interaction of electromagnetic waves with plasmas.

In an unmagnetized plasma the principal wave modes are, as discussed in the preceding sections, the transverse electromagnetic wave (T), the longitudinal electron plasma wave or Langmuir wave (L) and the low-frequency ion acoustic waves. Whether or not resonant three-wave interaction is possible depends on the simultaneous satisfaction of the resonance conditions and the dispersion relations.

Instabilities of particular interest and the electron densities at which they are expected are
(a) parametric decay instability (PD) at $N \lessapprox N_c$; the incident wave decays into an electron plasma wave and ion acoustic wave; $T \longrightarrow L + s$.
(b) Oscillating two-stream instability (OTS) at $N \simeq N_c$; a modified version of (a) where the ion acoustic wave is shifted to zero frequency.
(c) Two-plasmon decay (TPD) at $N \lessapprox N_c/4$; the incident wave decays into two Langmuir waves; $T \longrightarrow L + L'$.
(d) Stimulated Raman scattering (SRS) at $N < N_c/4$; the incident wave decays into another electromagnetic wave and an electron plasma wave; $T \longrightarrow T' + L$.
(e) Stimulated Brillouin scattering (SBS) at $N < N_c$; the incident wave decays into another electromagnetic wave and an ion acoustic wave; $T \longrightarrow T' + s$.

All of these processes may occur simultaneously in the plasma and may influence each other. If both the excited waves are purely electrostatic as in (a), (b) and (c) the decay process is absorptive in nature. The decay waves cannot leave the plasma and are ultimately absorbed which leads to an enhanced transfer of power from the incident wave to the plasma. If, however, one of the decay waves is electromagnetic as in (d), (e), it can escape from the plasma and the process appears as stimulated scattering of the incident wave.

From a quantum point of view, the three-wave interaction processes correspond to the decay of an incident photon into two other quanta such as photon + plasmon, photon + phonon or plasmon + plasmon. The frequency- and wavenumber matching conditions may then be interpreted as conservation of energy and momentum, respectively.

The thresholds for these instabilities in a homogeneous plasma are determined by the damping rates of the decay waves. In an inhomogeneous plasma the fact that the resonance conditions must be satisfied limits the generation of the instabilities to certain spatial regions even when the thresholds are exceeded. Propagation of the waves out of the resonant regions can limit the wave amplification and increase the threshold values over those in a homogeneous plasma.

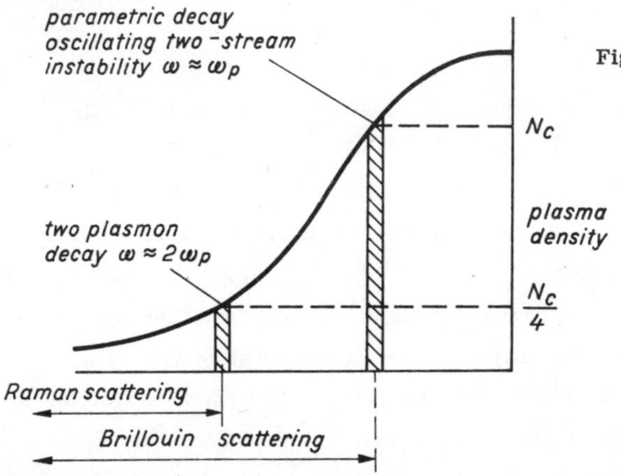

Fig. 7.1 Schematic diagram showing operation regions for different parametric instabilities when an electromagnetic wave is incident on a plasma with linear density gradient. The horizontal axis represents the distance along the plasma.

Fig. 7.1 shows in a schematic diagram at what density regions the different parametric instabilities occur in a nonuniform plasma.

In this section we confine ourselves to consider the two high-frequency decay processes SRS and TPD where only electron dynamics is involved. We shall first review the theory of SRS and TPD in a homogeneous plasma on the basis of mode coupling equations derived from the generalized nonlinear wave equation of Chapter 2. The mode coupling concept will thereafter be applied to an inhomogeneous plasma with a linear density profile without making use of the WKB approximation.

Conversely to harmonic generation, the analysis here cannot be started with the assumption of a stationary problem. This is due to the fact, that the plasma modes as partner for the interaction with the pump wave are of small amplitudes at the beginning and start growing from the thermal noise level. Thus, we have necessarily to do with a development in time which may or may not evolve into a stationary saturated final state.

7.2 SRS and TPD in a homogeneous plasma

The frequency matching condition (7.1) demands that both instabilities occur only at or below the plasma quarter critical density. From the dispersion relations (4.6), (4.7) it follows that both types of decay waves have frequencies above the plasma frequency. With $\omega_1, \omega_2 > \omega_p$ we have $\omega > 2\omega_p$ according to Eq. (7.1) in both cases which is equivalent to

$$N < N_c/4 \tag{7.3}$$

where N_c is the critical density associated with the pump frequency ω. This means that SRS and TPD cannot operate in a homogeneous plasma whose density exceeds the quarter critical density. In an inhomogeneous plasma the operation regimes of both instabilities are restricted as follows: Since the frequency of an electron plasma wave is confined to $\omega \lesssim \omega_p$ frequency matching for TPD can occur if $\omega \lesssim 2\omega_p$. This instability can therefore be expected to occur only near the quarter critical layer of a nonuniform plasma. The SRS instability, on the other hand, can occur in the whole region $N < N_c/4$ because the frequency of an electromagnetic wave is only bounded toward lower frequencies by the condition $\omega > \omega_p$.

The frequency matching conditions do not allow the direct generation of the frequency $3\omega/2$ in SRS or TPD. They can only be generated through secondary interaction of the Langmuir decay waves with the incident wave or with itself.

Both processes are of potential importance for laser-plasma interaction. Occurring in the underdense coronal region of the plasma, SRS may cause the plasma boundary to become reflecting and so preventing a certain fraction of the incident laser energy from being absorbed by the plasma. Due to its absorptive character, TPD, on the other hand, provides a mechanism of conversion of the incident radiation into energy of plasma particles via collisional absorption, Landau damping or particle trapping. Both processes are also responsible for the production of energetic electrons. This is because the Langmuir waves associated with both instabilities can have phase velocities of the order of the light velocity and therefore produce electrons of very high energy. Hot electron production via SRS and/or TPD has been observed and analyzed in various laser-plasma experiments /15, 16, 18, 21, 23, 25, 46 - 48/.

The mechanisms of stimulated Raman scattering and two-plasmon decay in homogeneous plasmas are well understood /33 - 45/. We shall give a brief review of the corresponding theory. We consider an infinite plasma of density N and represent the total electric field as a superposition of the fields associated with the three interacting waves

$$\vec{E}(\vec{r}, t) = \mathrm{re}\left\{ \vec{E}_o(t) e^{i(\vec{k}_o\vec{r} - \omega_o t)} + E_1(t) e^{i(\vec{k}_1\vec{r} - \omega_1 t)} + \vec{E}_2(t) e^{i(\vec{k}_2\vec{r} - \omega_2 t)} \right\}. \quad (7.4)$$

The indices 0, 1, 2 refer to the pump wave and the two decay waves, respectively. The electron velocity is represented in a similar way. We consider a space-independent interaction and allow the complex amplitudes $\vec{E}_j$, $\vec{v}_{ej}$ to depend only on time. They are further assumed to vary slowly on the time scale $1/\omega_j$. This approximation is known as weak coupling limit. That is, the waves preserve their identity as eigenmodes of the plasma and keep obeying one of the linear dispersion relations (4.6), (4.7) during the interaction. The pump wave is assumed to be a transverse wave satisfying the dispersion relation

$$k_o^2 c^2 = \omega_o^2 - \omega_p^2 .$$

The decay waves satisfy one of the two relations

$$k_T^2 c^2 = \omega_T^2 - \omega_p^2 ,$$

$$3\, k_L^2 v_T^2 = \omega_L^2 - \omega_p^2$$

according to their polarization.

When introducing the ansatz (7.4) into the generalized wave Eq. (2.12) and making use of the dispersion relations, the left-hand side takes the form

$$-\frac{\partial^2\vec{E}}{\partial t^2} - c^2 \text{ rotrot } \vec{E} + 3v_T^2 \text{ graddiv } \vec{E} - \omega_p^2 \vec{E}$$

$$= \text{Re}\left\{\sum_{j=o}^{2} 2i\omega_j \frac{d\vec{E}_j}{dt} e^{i(\vec{k}_j\vec{r} - \omega_j t)}\right\}.$$

Thereby, the terms $d^2\vec{E}_j/dt^2$ have been neglected against the terms $\omega_j \, d\vec{E}_j/dt$ according to the weak coupling approximation.

When substituting expression (7.4) into the nonlinear source term (2.13) on the right-hand side of the generalized wave equation, only terms are retained which vary proportional to $\exp i(\vec{k}_j\vec{r} - \omega_j t)$, $j = 0, 1, 2$ Selecting terms according to their exponentials yields finally the three-mode coupling equations

$$2i\omega_o \frac{d\vec{E}_o}{dt} = \frac{1}{2}\left\{i\frac{m_e}{e}\omega_p^2 \vec{k}_o(\vec{v}_{e1}\vec{v}_{e2}) + \omega_o\left[\vec{v}_{e1}(\vec{k}_2\vec{E}_2) + \vec{v}_{e2}(\vec{k}_1\vec{E}_1)\right]\right\}, \tag{7.5}$$

$$2i\omega_1 \frac{d\vec{E}_1}{dt} = \frac{1}{2}\left\{i\frac{m_e}{e}\omega_p^2 \vec{k}_1(\vec{v}_{eo}\vec{v}^{\,*}_{e2}) + \omega_1\left[-\vec{v}_{eo}(\vec{k}_2\vec{E}^{*}_2) + \vec{v}^{\,*}_{e2}(\vec{k}_o\vec{E}_o)\right]\right\}, \tag{7.6}$$

$$2i\omega_2 \frac{d\vec{E}_2}{dt} = \frac{1}{2}\left\{i\frac{m_e}{e}\omega_p^2 \vec{k}_2(\vec{v}_{eo}\vec{v}^{\,*}_{e1}) + \omega_2\left[-\vec{v}_{eo}(\vec{k}_1\vec{E}^{*}_1) + \vec{v}^{\,*}_{e1}(\vec{k}_o\vec{E}_o)\right]\right\}, \tag{7.7}$$

where collisional damping has been ignored. The electron velocity can be eliminated with the help of the simplified electron momentum equation

$$\vec{v}_{ej} = \frac{e\vec{E}_j}{i\, m_e \omega_j}.$$

This results in a closed system for the amplitudes $\vec{E}_j(t)$.

Parametric approximation

To study the initial stage of the instability it is appropriate to ignore the change in the pump amplitude and to treat it as a constant parameter. This simplifies the analysis in that the two remaining equations become linear

Stimulated Raman scattering: We specify the decay wave 1 to be longitudinal and the decay wave 2 to be transverse and denote the polarization by the subscript L or T. A sketch of the geometry is shown in Fig 7.2 (a). Denoting by $\vec{e}_o$, $\vec{e}_T$ unit vectors in direction of $\vec{E}_o$, $\vec{E}_T$ we get after scalar multiplication of Eq. (7.6) by $\vec{k}_L$ and Eq. (7.7) by $\vec{e}_T$ the pair of equations

$$\frac{dn_L}{dt} = -\frac{1}{4i} v_E k_L^2 \frac{\omega_p^2}{\omega_L \omega_T} (\vec{e}_o\vec{e}_T) E^{*}_T, \tag{7.8}$$

$$\frac{dE_T}{dt} = -\frac{1}{4i} v_E (\vec{e}_o\vec{e}_T) n^{*}_L \tag{7.9}$$

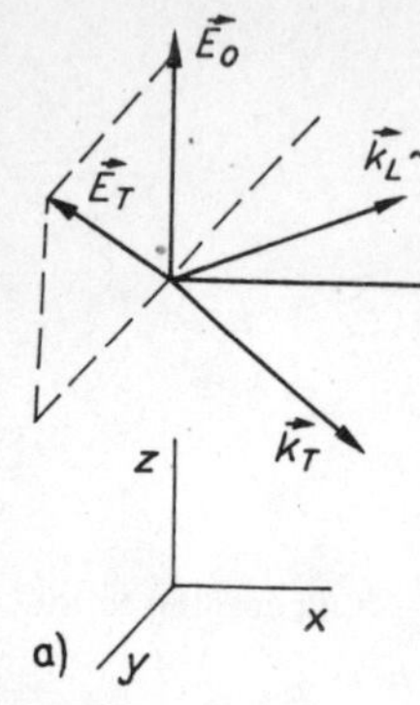

Fig. 7.2 (a) Vector constellation at stimulated Raman scattering

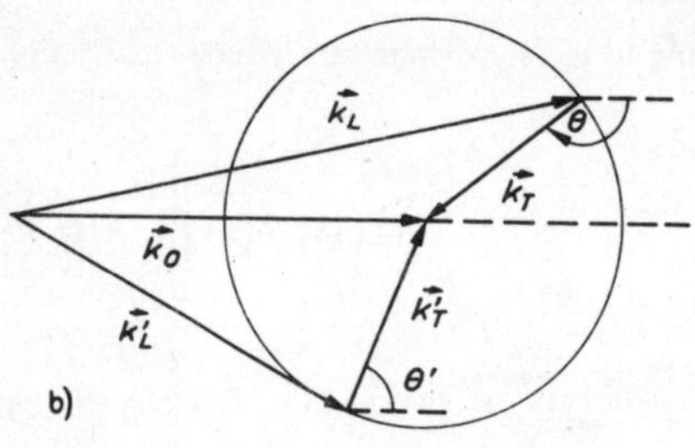

Fig. 7.2 (b) Two different vector configurations $\vec{k}_o$, $\vec{k}_T$, $\vec{k}_L$ for fixed scattering frequency at stimulated Raman scattering, illustrating the influence of the scattering angle 0 on the magnetide of k_L, i. e., on the growth rate, according to Eq. (7.10)

where $n_L = (\vec{k}_L \vec{E}_L)$, $\vec{E}_T = \vec{e}_T E_T$ and $v_E = |\vec{v}_{eo}| = e E_o/m_e \omega_o$. Putting n_L, $E_T \propto \exp(\gamma t)$ we get from Eqs. (7.8), (7.9) the growth rate

$$\gamma_{SRS} = \frac{k_L v_E}{4} \frac{\omega_p}{\sqrt{\omega_L \omega_T}} (\vec{e}_o \vec{e}_T) \simeq \frac{k_L v_E}{4} \left(\frac{\omega_p}{\omega_T} \right)^{1/2} (\vec{e}_o \vec{e}_T) . \tag{7.10}$$

Obviously the SRS instability is driven most efficiently when the polarization directions of pump wave and scattered wave coincide. The scattering angle does not enter Eq. (7.10) directly. It influences the growth rate via the frequency- and wavenumber matching relations which determined $\vec{k}_L$ and ω_T. From the dispersion relation of the scattered wave

$$c^2 k_T^2 = \omega_T^2 - \omega_p^2$$

we find, using $\vec{k}_T = \vec{k}_o - \vec{k}_L$ and the dispersion relation $c^2 k_o^2 = \omega_o^2 - \omega_p^2$ for the pump wave, that

$$\vec{k}_L (2 \vec{k}_o - \vec{k}_L) = (\vec{k}_o - \vec{k}_T)(\vec{k}_o + \vec{k}_T) = \frac{1}{c^2} (\omega_o^2 - \omega_T^2) .$$

Fig. 7.2 (b) illustrates the associated vector constellations of $\vec{k}_o$, $\vec{k}_L$ and $\vec{k}_T$ for fixed scattering frequency. The top of $\vec{k}_L$ can vary along the circle shown in the Figure. As may be seen, backscattering produces the largest value of k_L, i. e., the maximum growth rate. SRS occurs over the whole range of scattering angles, but when a plane pump wave is normally incident upon a plane-stratified plasma density profile, the transverse directions are ignorable and it is sufficient to consider only forward and backward scattering.

Fig. 7.3 shows the growth rate γ_{SRS} in the cases of forward and backward scattering as function of the plasma density for a background temperature of $T_e = 1$ keV ($v_T/c \simeq 0.05$). The normalized quantity

$$\Gamma = \frac{\gamma_{SRS}/\omega_o}{v_E/c}$$

is plotted which is apparently independent of the pump intensity. As already mentioned, stimulated Raman scattering operates in the whole density region below quarter critical density. As seen from Fig. 7.3, backscattering represents the most unstable geometry. Except the vicinity of $N_c/4$, the growth rate of forward scattering is several times less than for backscattering. Note the relatively weak dependence of the normalized growth rate on the plasma density. At the quarter critical density the curves for back and forward scattering meet. In addition, Figs. 7.3 (b), (c) show the normalized scattering frequency ω_T/ω_o and the quantity $k_L \lambda_D$ as function of the plasma density. Clearly, for low plasma densities the down shift of the scattered frequency relative to ω_o is small ($\omega_T \lesssim \omega_o$); with increasing density the shift increases and ω_T approaches $\omega_o/2$. Close to the quarter critical density the scattered frequency is shifted even slightly below $\omega_o/2$.

The quantity $k_L \lambda_D$ is a measure for the phase velocity of the Langmuir wave and indicates the influence of Landau damping. While the wave corresponding to forward scattering has generally a phase velocity close to the light velocity, the wave corresponding to backscattering has

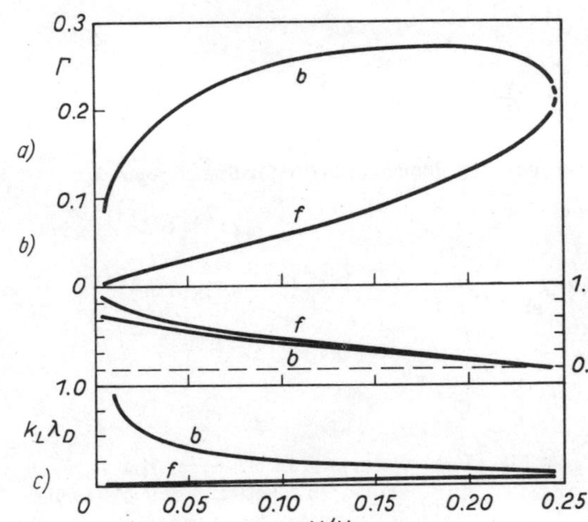

Fig. 7.3 Normalized growth rate $\Gamma = (\gamma_{SRS}/\omega_o)/(v_E/c)$ (a), relative scattering frequency ω_T/ω_o (b) and $k_L \lambda_D$ (c) as function of the plasma density for stimulated Raman scattering in a homogeneous plasma ($v_T/c = 0.05$)

a phase velocity much less than c. Therefore, backscattering is sensitive to Landau damping while forward scattering is practically not affected. Calculations for different electron temperatures (ignoring Landau damping) indicate only little influence of T_e on the growth rate. Only for extremely high temperatures (beyond $v_T/c = 0,2$, e. g., $T_e \simeq 20$ keV) a decrease sets in. With growing temperature, however, Landau damping in the case of backscattering becomes important which can significantly restrict the density region over which Raman backscattering operates. As simulations have shown /41, 42/, at a temperature of $T_e = 20$ keV the plasma waves associated with backscattering have such short wavelengths that Landau damping dominates at any plasma density and Raman backscattering will be excluded. The stimulated scattering process will then be dominated by forward scattering. In addition, growing temperature enlarges the density region near $N_c/4$ where the scattering frequency ω_T is shifted below $\omega_0/2$.

Fig. 7.4 completes the presentation showing a polar diagram for the normalized growth rate as function of the scattering angle relative to the propagation direction of the pump wave for two plasma densities.

When collisional and Landau damping is taken into consideration the mode coupling equations predict a pump intensity threshold $v_{E,\,thr}$ for the SRS instability to be excited. According to Eq. (A 12) of the appendix it depends on the coupling coefficients c_1, c_2 and the damping decrements σ_1, σ_2 as

$$v_{E,\,thr.} = \left(\frac{\sigma_1 \sigma_2}{c_1 c_2^*} \right)^{1/2}$$

Introducing the coupling coefficients

$$c_T = -\frac{k_L^2}{4i} \frac{\omega_p^2}{\omega_L \omega_T} , \quad c_L = -\frac{1}{4i} ,$$

defined by Eqs. (7.8), (7.9) and the temporal damping decrements according to formulas (4.10), (4.16)

$$\sigma_T = \frac{\omega_p^2}{\omega_T^2} \frac{\nu_{ei}}{2}, \quad \sigma_L = \frac{1}{2} (\nu_{ei} + \nu_L^*)$$

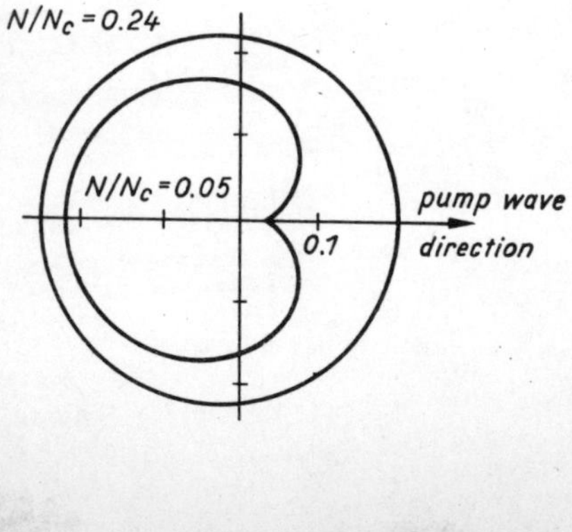

Fig. 7.4 Polar plot of the normalized growth rate Γ as function of the scattering angle for two plasma densities

the threshold intensity is given by

$$v_{E,\,thr}^{SRS} = \frac{2}{k_L}\left\{\frac{\omega_p}{\omega_T}\,\nu_{ei}\,(\nu_{ei}+\nu_L^{*})\right\}^{1/2} .$$

Electron-ion-collisions can become important with increasing pump frequencies in suppressing SRS by absorbing the incident radiation, by damping of the electron plasma waves in addition to Landau damping and by absorbing the scattered radiation that contributes to the amplification /44, 49/.

Two-plasmon decay: In this case both decay waves are longitudinal. The process is two-dimensional because the matching conditions and the dispersion relations cannot be satisfied simultaneously in only one dimension. The geometry is sketched in Fig. 7.5. Scalar multiplication of Eq. (7.6) by k_L and of Eq. (7.7) by $k_{L'}$ results in the two equations

$$\frac{dn_L}{dt} = -\frac{k_L v_E}{4i}(\vec{e}_o\vec{e}_L)\left\{1-\Omega\, k_L^2/k_{L'}^2\right\} n_{L'}^{*}, \qquad (7.11)$$

$$\frac{dn_{L'}}{dt} = -\frac{k_L v_E}{4i}(\vec{e}_o\vec{e}_L)\left\{\Omega\, k_{L'}^2/k_L^2-1\right\} n_L^{*} \qquad (7.12)$$

where $n_L = (\vec{k}_L\vec{E}_L)$, $n_{L'} = (\vec{k}_{L'}\vec{E}_{L'})$ and $\Omega = \omega_p^2/\omega_L\,\omega_{L'}$. In deriving these equations the relation

$$(\vec{e}_o\vec{e}_L) = -k_L/k_{L'}\,(\vec{e}_o\vec{e}_{L'})$$

has been used which follows from $(\vec{k}_o\vec{e}_o) = 0$. Combining Eqs. (7.11), (7.12) and looking for solution $\propto \exp\gamma t$ yields the growth rate for the two-plasmon decay

$$\gamma_{TPD} = \frac{k_L v_{eo}}{4}(\vec{e}_o\vec{e}_L)\left[\Omega\left\{(k_L/k_{L'})^2+(k_{L'}/k_L)^2\right\}-\Omega^2-1\right]^{1/2} . \qquad (7.13)$$

Only little information can directly be gained from this expression because ω_L, $\omega_{L'}$, $\vec{k}_L$, $\vec{k}_{L'}$ are determined by the frequency- and wavenumber matching relations together with the dispersion laws. It is, however, immediately seen from Eq. (7.13) that the growth rate vanishes

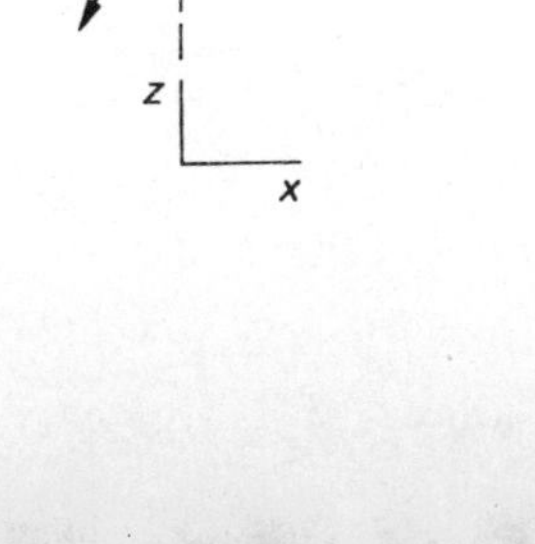

Fig. 7.5 Vector constellation at two-plasmon decay

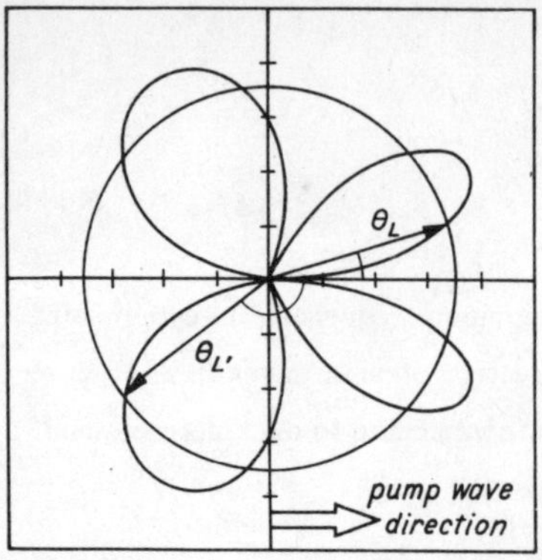

Fig. 7.6 Polar diagram of the normalized growth rate Γ for two-plasmon decay as function of the angle $\sphericalangle\ (\vec{k}_L, \vec{k}_o)$ between the pump wave direction and the propagation direction of one of the longitudinal decay waves. As an example it is shown what pair of angles θ_L, $\theta_{L'}$ belongs to a certain growth rate Γ, $0 < \Gamma < \Gamma_{max}$.

when the longitudinal decay waves are polarized in the plane perpendicular to $\vec{E}_o$ owing to the factor $(\vec{e}_o\,\vec{e}_L)$. Consequently, decay wave propagating parallel or antiparallel to $\vec{k}_o$ cannot grow as it is the case in stimulated Raman scattering. It can further be concluded from Eq. (7.13) that no unstabil behaviour occurs for equal wavenumbers $k_L = k_{L'}$, that is, for equal frequencies of the two Langmuir waves. In this case, the argument of the square root takes the form

$$2\Omega - \Omega^2 - 1 = -(1-\Omega)^2 \leq 0 .$$

That is, Eq. (7.13) does not admit a real positive growth rate. This case is particularly realized, when the pump field is treated in dipole approximation (long-wavelength approximation, $\vec{k}_o = 0$). Fig. 7.6 tries to illustrate what Eq. (7.13) predicts in combination with the resonance conditions and the dispersion relations concerning possible vector constallations in TPD and the associated growth rates. The $\vec{k}$-vectors of the decay waves are taken to vary in the $\vec{E}_o - \vec{k}_o$-plane (otherwise the growth rate will be smaller by the factor $\cos \emptyset$ where $\emptyset$ is the angle between the $\vec{k}$-vector and the $\vec{E}_o - \vec{k}_o$-plane). It is demonstrated in this figure how a pair of angles θ, θ' can be constructed for a certain normalized growth rate. (Note that the vectors shown in Fig. 7.6 have the directions of the $\vec{k}$-vectors but their length corresponds to the growth rate.) Fig. 7.7 shows in addition a sequence of such angular diagrams for different combinations of plasma density and temperature. It is seen from this figures that generally the optimum growth

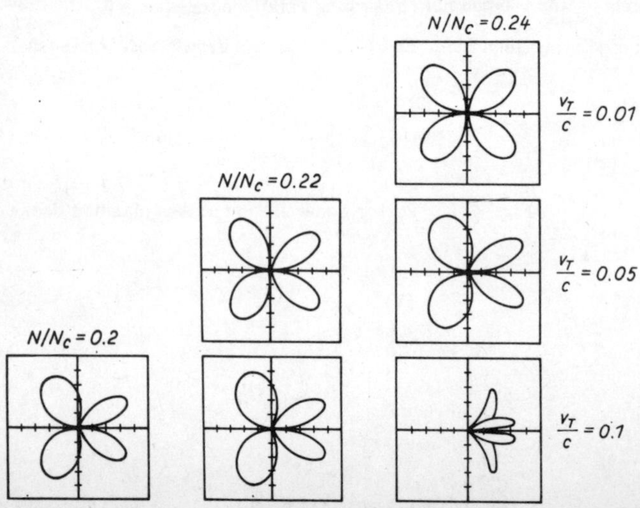

Fig. 7.7 Polar plots of the TPD growth rate (as Fig. 7.6) for different densities and temperatures

rate is attained when the wave vector of one of the decay waves almost bisects the angle between $\vec{k}_o$ and $\vec{E}_o$, i. e., for pairs of angles near $\theta_L = \pm 45^o$, $\theta_{L'} = \pm 135^o$. This symmetric configuration occurs preferentially for low temperatures and densities not too close to $N_c/4$. With growing electron temperature the symmetry is reduced and at sufficiently hot plasmas a finite growth rate is possible at densities near the quarter critical density even for decay waves having both a positive component of their wave vectors along the pump wave direction. The dominant role of the symmetric arrangement of the $\vec{k}$-vectors of the decay waves at TPD has been confirmed several times in experiments performed in almost homogeneous plasmas /15, 25/.

Fig. 7.8 completes the picture, illustrating the variation of the optimum growth rate for TPD with density for three different electron temperatures. The normalized growth rate $\Gamma = (\gamma_{TPD}/\omega_o)/(v_E/c)$ attains a value slightly above 0.2 as maximum. A growth rate $\Gamma = 0.25$ which corresponds to $\gamma_{TPD} = k_o v_e/4$ and which is often refered to as maximum growth rate for TPD in the literature is apparently not reached. It is once more seen from Fig. 7.8 that the resonance conditions impose stringent restrictions on the range of densities over which this instability can be excited, in contrast to stimulated Raman scattering.

Nonlinear stage of SRS and TPD

The parametric approximation which predicts an exponential growth of the decay wave amplitudes at a linear growth rate breaks down when the latter ones become comparable to the pump amplitude or when the energy fed into the decay waves is dissipated nonlinearly. In the first case, mode coupling reduces the pump amplitude and simultaneously the growth rate In the damping-free limit the long-time behaviour of SRS, for instance, is cyclic as energy is transferred back and forth between the pump wave and the decay waves in a periodic sequence of a Stokes- and an anti-Stokes scattering process. Initially the decay wave amplitudes grow exponentially according to the prediction of the parametric theory (Stokes process) and reach maxima at the same time as the pump wave is completely depleted. Subsequently the anti-Stokes process startes during which the pump wave regains its previous maximum. When damping is included the system evolves into a steady state with finite levels of the decay wave amplitudes and completely depleted pump wave For details we refer to /42/, where such calculations have been carried out or to similar calculations performed in the case of stimulated Brillouin-

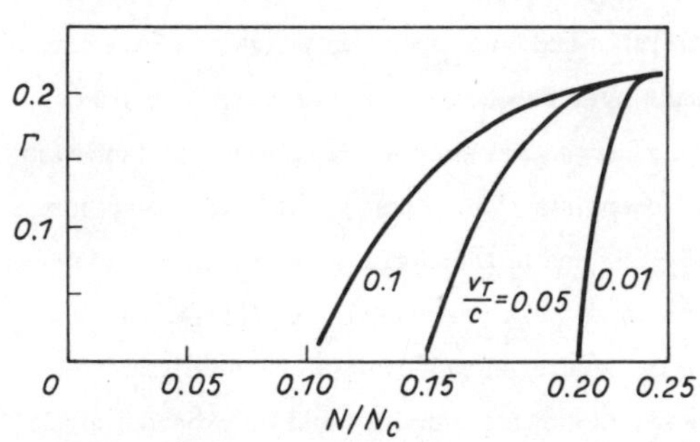

Fig. 7.8 Maximum TPD growth rates versus plasma density for three temperatures

scattering in a homogeneous unbounded plasma, presented in Chapter 9.

Particle simulations of SRS in a homogeneous plasma /42, 50/ demonstrated, however, that mode coupling effects like pump depletion dominate only for small pump intensities not far above threshold as nonlinear saturation mechanism, namely, as long as the Langmuir wave amplitude does not exceed the critical value for electron trapping. When, however, the plasma waves grow to an amplitude sufficiently large to interact strongly with the particles, their damping increases drastically since the waves begin to strongly accelerate large numbers of electrons from the main body of the plasma. That is, an efficient energy transfer to the particles takes place. The critical Langmuir wave amplitude for electron trapping to occur is approximately given by the relation

$$\left(\frac{2eE_L}{m_e k_L}\right)^{1/2} \gtrsim \frac{\omega_L}{k_L} - 2v_T \,. \tag{7.14}$$

At an electron temperature of $T_e = 1$ keV ($v_T/c \simeq 0.05$) this critical amplitude is $v_E/c \simeq 0.15$. Electron trapping considerably reduces the growth rate and leads to the formation of a non-thermal energetic tail in the electron velocity distribution function on the expense of the pump wave energy. This mechanism can only be phenomenologically included in the fluid description.

Saturation of SRS and TPD due to secondary decay of the generated Langmuir waves in another Langmuir wave and an ion acoustic wave has been studied in /51/.

7.3 SRS and TPD in an inhomogeneous plasma

The presence of a density gradient significantly modifies the features of stimulated Raman scattering and two-plasmon decay. Since the wave numbers of the natural plasma modes depend on density, zero-order density gradients limit the spatial region over which resonant interaction occurs for a given set of three waves. In this sense a density gradient has a similar effect on these instabilities as confining the interaction region in a homogeneous plasma to a finite length. Propagation of wave energy out of the interaction region represents an effective damping on each wave and can lead to reduced growth rates and correspondingly higher thresholds. The convection-determined damping may dominate over that caused by dissipative processes, which set the thresholds in a homogeneous plasma. A general study of parametric instabilities in an inhomogeneous plasma was first given by Rosenbluth /52/. Subsequently various authors contributed to the corresponding theory /41, 53 - 69/ and in addition a number of numerical simulations of SRS and TPD in inhomogeneous plasmas were performed /41, 42, 44, 70 - 75/.

In contrast to a homogeneous plasma, where the scattered radiation is essentially monochromatic, in an inhomogeneous plasma always a range of frequencies must be expected for a fixed pump frequency and a fixed observation angle relative to the pump wave direction /64/.

This is due to the fact that the frequency- and wavenumber matching conditions may simultaneously be satisfied for different scattering frequencies at different portions of the density profile. In addition, the angle at which the scattered radiation is emitted from the plasma does in general not coincide with that angle at which it is generated in the interior of the plasma as a consequence of refraction on its way through the medium.

Another important question, at least from a theoretical point of view, which arises when considering an inhomogeneous or spatially limited plasma is whether an instability is absolute or convective /76, 77/. A convective instability usually saturates by linear processes as convection of wave energy out of the resonance region and the decay wave amplitudes grow only to a finite level in the linear parametric approximation (constant pump intensity). In an absolute instability, however, the decay wave amplitudes continue to grow in time. Saturation can only be provided by nonlinear processes and would, in general, be expected to take place at a higher level than in the convective case. Furthermore, absolute instabilities tend to be localized in space and can therefore locally perturb the plasma parameters in a very significant manner.

As far as a homogeneous medium of finite extent is concerned, Gorbunov /78/ showed, that oppositely directed components of the decay wave group velocities along the direction of the pump wave propagation are necessary for an absolute instability to occur. Hence, Raman backscattering and two-plasmon decay are candidates for absolute instabilities in an inhomogeneous plasma.

Another problem which complicates the situation arises due to the existence of turning points of the decay waves and the pump wave in an inhomogeneous plasma in which the density increases monotonically. Raman forward scattering and backscattering, for instance, may no longer be considered as well separated phenomena. Interference occurs via the following feedback loop: for fixed frequencies of the scattered waves both processes are resonant at two different positions upon a monotonic density profile. Waves amplified in one scattering region can provide via reflection the level from which growth startes in the other scattering region. It was first pointed out by Cairns /61/ (see also /67/) that this mechanism can establish an absolute instability that has no direct analog in a homogeneous plasma of finite extent.

In its general study on parametric three-wave interaction in an inhomogeneous medium Rosen senbluth /52/ showed first that no absolute instability can exist in an inhomogeneous plasma, when the wave number mismatch varies linearly in the neighbourhood of the resonance. The effect of the instability is then to produce spatial amplification of the waves, as they travel through the resonance region. The maximum amplification from a thermal source amplitude A_o is given by

$$A = A_o \exp\left\{ \pi \gamma_o^2 / |v_1 v_2 k'| \right\} \tag{7.15}$$

where γ_o is the growth rate for a homogeneous plasma and v_1, v_2 are the group velocities along the density gradient. Further,

$$k(x) = k_o(x) - k_1(x) - k_2(x) \tag{7.16}$$

is the wave number mismatch along the direction of propagation and

$$k'(x) = \frac{d}{dx}\left\{k_o(x) - k_1(x) - k_2(x)\right\} \tag{7.17}$$

is its spatial rate of change. For an effective growth according to Eq. (7.15), the condition

$$\pi\gamma_o^2 / |v_1 v_2 k'| > 1 \tag{7.18}$$

must be satisfied. That is, the wave must grow substantially during the time it propagates to the point where the phase mismatch is substantial. This greatly raises the threshold power needed for Raman backscattering for incidence normal to the density gradient because of the fast propagation of the backscattered radiation out of the interaction region. However, for the scattered radiation propagating almost perpendicular to the density gradient, its convection out of the resonant region along the density gradient is slow and a lower threshold for the side-scattering is thus expected /53, 68/.

An absolute instability in an inhomogeneous plasma can take place only when both the wave-number mismatch $k(x)$ and its rate of change $k'(x)$ along the direction of propagation vanishes simultaneously at a certain point of the density profile. This is indeed possible, for example, for Raman backscattering near the quarter critical density where the frequency of the incident wave is twice the local plasma frequency. These conclusions are the results of an analysis based on the WKB approximation. The decay waves, however, do not have a well defined wave-number in the vicinity of the point of perfect matching because it is close to their turning points. Thus the criterion for the validity of the WKB approximation is violated there. A detailed analysis of White et al. /56/ which avoids the WKB approximation confirmed the existence of an absolute Raman instability near $N_c/4$, indicating an extended domain of validity of the WKB theory. The same result has been found in a number of subsequent analytical and numerical studies.

The gradient-induced threshold for the absolute Raman backscatter instability near $N_c/4$ has been calculated to be /53/

$$\left(\frac{v_E}{c}\right)_{thr} = 2\,(k_o L)^{-2/3} \tag{7.19}$$

(k_o local pump wave number, close to the free space wave number of the pump, L density gradient scale length). Similarly, the threshold for the two-plasmon decay has been found to be /58/

$$\left(\frac{v_E}{c}\right)_{thr} = 2\sqrt{3}\,\frac{v_T}{c}\,(k_o L)^{-1/2}\,. \tag{7.20}$$

Note that these expressions differ from those given in the original paper by a factor 2. This is due to the fact that the real pump field in /53, 58/ is represented as $E_o \exp i(k_o x - \omega_o t)$ + cc.. In this study, however, we use the representation $\mathrm{Re}\left\{E_o \exp i(k_o x - \omega_o t)\right\} = 1/2\ \times$

$\{ E_o \exp i(k_o x - \omega_o t) + cc. \}$ which results in the additional factor 2 mentioned above. Because of the factor v_T/c in Eq. (7.20) the two-plasmon decay appears as less sensitive to a density gradient than Raman scattering.

The basic idea behind the analysis which yields the above thresholds is an estimation of the convection-dominated damping on the decay waves as

$$\sigma_{1,2} = V_{1,2}/l$$

where $V_{1,2}$ are the group velocities and l is the size of the interaction regime. l is estimated by the condition that the integral over the wave number mismatch with respect to x over the distance l becomes one half.

7.4 Mode coupling equations for SRS and TPD in an inhomogeneous plasma

In what follows we derive mode coupling equations for stimulated Raman scattering and two-plasmon decay in a nonuniform plasma from the basic equation for high-frequency processes and solve them numerically without making use of the WKB approximation. Thereby we concentrate on the initial parametric stage of the absolute instabilities and look for predictions that can be made from the hydrodynamic model concerning the dependence of thresholds and growth rates on the relevant physical parameters /65/.

We consider a one-dimensionally layered medium and assume the plasma density to vary along the x-direction. Starting point is again the generalized wave Eq. (2.12). Similarly to the homogeneous case we represent the total electric field as superposition of the fields associated with the three interacting waves

$$\vec{E}(\vec{r}, t) = \sum_{j=0}^{2} \vec{E}_j(\vec{r}, t) = \mathrm{Re}\left\{ \sum_{j=0}^{2} \vec{E}_j(x, t)\, e^{i(k_z z - \omega_j t)} \right\}. \tag{7.21}$$

Conversely to ansatz (4.6), no explicit assumption is imposed here for the spatial variation of the complex amplitudes $E_j(x, t)$, that is, no use is made of the WKB approximation. The variation in time, however, is again assumed to be slow on the time scale $1/\omega_j$. To simplify the analysis we restrict ourselves to a pump wave propagating along the density gradient (i. e., in x-direction). We then can write

$$\vec{E}_o(x, t) = E_o(x, t)\, \vec{e}_z , \quad k_{oz} = 0 . \tag{7.22}$$

As far as stimulated Raman scattering is concerned we further assume that also the decay waves propagate along the x-direction. This implies that only forward- and backward scattering is included. The longitudinal and the transverse decay wave are decoupled and we may write

$$\vec{E}_1(x, t) = E_L(x, t)\,\vec{e}_x, \qquad k_{1z} = 0, \tag{7.23}$$

$$\vec{E}_2(x, t) = E_T(x, t)\,\vec{e}_z, \qquad k_{2z} = 0. \tag{7.24}$$

Only the frequency matching condition

$$\omega_o = \omega_L + \omega_T$$

has to be taken into consideration here. Introducing the total field (7.21) together with the specifications (7.22) - (7.24) into the generalized wave equation (2.12) and selecting terms according to their frequency dependence yields the following three coupled equations

$$2i\omega_o \frac{\partial E_o}{\partial t} + c^2 \frac{\partial^2 E_o}{\partial x^2} + \omega_o^2\, \varepsilon(\omega_o, N)\, E_o = -\frac{e}{2m_e}\frac{\omega_o}{\omega_T} E_T \frac{\partial E_L}{\partial x}, \tag{7.25}$$

$$2i\omega_L \frac{\partial E_L}{\partial t} + 3v_T^2 \frac{\partial^2 E_L}{\partial x^2} + \omega_L^2\, \varepsilon(\omega_L, N)\, E_L = \frac{e}{2m_e}\frac{\omega_p^2}{\omega_L\,\omega_o} \frac{\partial}{\partial x}(E_o E_T^*), \tag{7.26}$$

$$2i\omega_T \frac{\partial E_T}{\partial t} + c^2 \frac{\partial^2 E_T}{\partial x^2} + \omega_T^2\, \varepsilon(\omega_T, N)\, E_T = -\frac{e}{2m_e}\frac{\omega_T}{\omega_o} E_o \frac{\partial E_L^*}{\partial x}. \tag{7.27}$$

This set of equations describes the combined operation of stimulated Raman forward- and backscattering in an onedimensionally nonuniform plasma when the pump wave is normally incident. We point again to the fact that no restriction is imposed on the spatial variation of E_o, E_L, E_T. Landau damping of the longitudinal decay wave can phenomenologically be included by replacing v_T^2 according to Eq. (4.15).

The subsequent interaction of the longitudinal wave with the pump wave to produce a transverse wave with the frequency

$$\omega_{TH} = \omega_L + \omega_o$$

can simultaneously be described by the additional equation

$$2i\omega_{TH} \frac{\partial E_{TH}}{\partial t} + c^2 \frac{\partial^2 E_{TH}}{\partial x^2} + \omega_{TH}^2\, \varepsilon(\omega_{TH}, N)\, E_{TH} = -\frac{e}{2m_e}\frac{\omega_{TH}}{\omega_o} E_o \frac{\partial E_L}{\partial x} \tag{7.28}$$

This process, however, has no influence on the stimulated scattering process.

As already pointed out, the two-plasmon decay cannot be reduced to an one-dimensional problem. We assume the pump wave again to propagate along the density gradient, the two decay waves, however, must have finite wavenumbers k_{1z}, $k_{2z} \neq 0$ in transverse direction for TPD to operate. We represent the two p-polarized decay wave amplitudes $\vec{E}_1(x, t)$, $\vec{E}_2(x, t)$ as

$$\vec{E}_j(x, t) = E_{jx}(x, t)\,\vec{e}_x + E_{jz}(x, t)\,\vec{e}_z \qquad j = 1, 2 . \tag{7.29}$$

Owing to $k_{oz} = 0$ the resonance conditions reduce to

$$\omega_o = \omega_1 + \omega_2 , \qquad 0 = k_{1z} + k_{2z} . \tag{7.30}$$

When introducing ansatz (7.21) with the specified expressions (7.22) and (7.29) into the generalized wave Eq. (2.12) we get after frequency selection the following mode coupling system

$$2i\omega_o \frac{\partial E_o}{\partial t} + c^2 \frac{\partial^2 E_o}{\partial x^2} + \omega_o^2\, \mathcal{E}(\omega_o, N)\, E_o =$$

$$= -\frac{e}{2m_e}\left\{\frac{\omega_o}{\omega_1} E_{1z}\left(\frac{\partial E_{2x}}{\partial x} + ik_{2z}E_{2z}\right) + \frac{\omega_o}{\omega_2} E_{2z}\left(\frac{\partial E_{1x}}{\partial x} + ik_{1z}E_{1z}\right)\right\} \tag{7.31}$$

(equation for the pump wave);

$$2i\omega_1 \frac{\partial E_{1x}}{\partial t} + 3v_T^2 \frac{\partial^2 E_{1x}}{\partial x^2} + \omega_1^2\left\{\mathcal{E}(\omega_1, N) - \frac{k_{1z}^2}{k_1^2}\right\} E_{1x} -$$

$$- ik_{1z}(c^2 - 3v_T^2)\frac{\partial E_{1z}}{\partial x} = \frac{e}{2m_e}\frac{\omega_p^2}{\omega_o\omega_2}\frac{\partial}{\partial x}(E_o E_{2z}^*) , \tag{7.32}$$

$$2i\omega_1 \frac{\partial E_{1z}}{\partial t} + c^2 \frac{\partial^2 E_{1z}}{\partial x^2} + \omega_1^2\left\{\mathcal{E}(\omega_1, N) - \frac{3v_T^2}{c^2}\frac{k_{1z}^2}{k_1^2}\right\} E_{1z} -$$

$$- ik_{1z}(c^2 - 3v_T^2)\frac{\partial E_{1z}}{\partial x} = -i\frac{e}{2m_e}\frac{\omega_1}{\omega_o} E_o\left\{\left(\frac{\omega_p^2}{\omega_1\omega_2}k_{1z} + k_{2z}\right)E_{2z}^* + i\frac{\partial E_{2x}^*}{\partial x}\right\} \tag{7.33}$$

(equations for the x- and z-component of the first decay wave);

$$2i\omega_2 \frac{\partial E_{2x}}{\partial t} + 3v_T^2 \frac{\partial^2 E_{2x}}{\partial x^2} + \omega_2^2\left\{\mathcal{E}(\omega_2, N) - \frac{k_{2z}^2}{k_2^2}\right\} E_{2x} -$$

$$- ik_{2z}(c^2 - 3v_T^2)\frac{\partial E_{2x}}{\partial x} = \frac{e}{2m_e}\frac{\omega_p^2}{\omega_o\omega_1}\frac{\partial}{\partial x}(E_o E_{1z}^*) , \tag{7.34}$$

$$2i\omega_2 \frac{\partial E_{2z}}{\partial t} + c^2 \frac{\partial^2 E_{2z}}{\partial x^2} + \omega_2^2\left\{\mathcal{E}(\omega_2, N) - \frac{3v_T^2}{c^2}\frac{k_{2z}^2}{k_2^2}\right\} E_{2z} -$$

$$- ik_{z2}(c^2 - 3v_T^2)\frac{\partial E_{2x}}{\partial x} = - i\frac{e}{2m_e}\frac{\omega_2}{\omega_o}E_o\left[\left(\frac{\omega_p^2}{\omega_1\omega_2}k_{2z} + k_{1z}\right)E^*_{1z} + i\frac{\partial E^*_{1x}}{\partial x}\right] \tag{7.35}$$

(the same for the second decay wave). The denotation $k_j = c/\omega_j$ has been used. This system, which appears to be a little more complicated than that for SRS, describes two-plasmon decay in a plasma whose density varies in x-direction for a normally incident pump wave. The transverse wave numbers k_{1z}, k_{2z} appear as additional parameters, linked by the resonance condition ($k_{2z} = - k_{1z}$).

Both mode coupling systems have been solved numerically according to the following scheme: a linearly increasing density ramp has been assumed extending from zero up to a density fairly above quarter critical. Pump wave reflection can be simulated by either chosing the uniform density adjacent to the linearly increasing part as overcritical or by letting the rising part of the profile extend beyond the critical density. In the latter case, however, the space interval for computation covers uninteresting parts between quarter critical and critical density, where essentially only the pump wave is present.

To initiate the stimulated decay process a non-zero initial level of at least one of the decay wave amplitudes is necessary. This implies that the initial frequencies of the decay waves must be fixed. This, however, does not crucially influence the time development of the instabilities since the time derivatives in the mode coupling equations allow changes in frequencies which may develop self-consistently toward steady values.

7.5. Results of mode coupling analysis of SRS, TPD

The mode coupling equations derived in the previous section are controlled by three parameters: (a) the pump amplitude, represented by the parameter $v_E/c = eE_o/m_e\omega_o c$, (b) the ratio L/λ_o of the density gradient scale length to the free space pump wave length, and (c) the electron temperature, represented by the parameter $v_T/c = (k_B T_e/m_e)^{1/2}/c$. The two sets of mode coupling equations describing SRS and TPD have been solved for a variety of combinations of the three parameters v_E/c, L/λ_o, v_T/c. Typical results are illustrated in the following figures.

Stimulated Raman Scattering: Fig. 7.9 shows the temporal growth of the maximum amplitudes of the decay waves within the plasma and in addition the free space amplitudes of both the backscattered wave and the $3\omega_o/2$-wave. Parameters are $v_E/c = 0.1$, $L = 50\lambda_o$, $v_T/c = 0.05$. The absolute character of the SRS instability for this choice of parameters is apparent, indicated by the unlimited exponential growth of $|E_L|_{max}$ and $|E_T|_{max}$ at a linear growth rate as long as pump depletion can be ignored. The frequencies of the decay waves tend toward values near half the pump frequency regardless of their initial values. This implies that the instability concentrates at a portion of the density profile slightly below the quarter critical density. This is

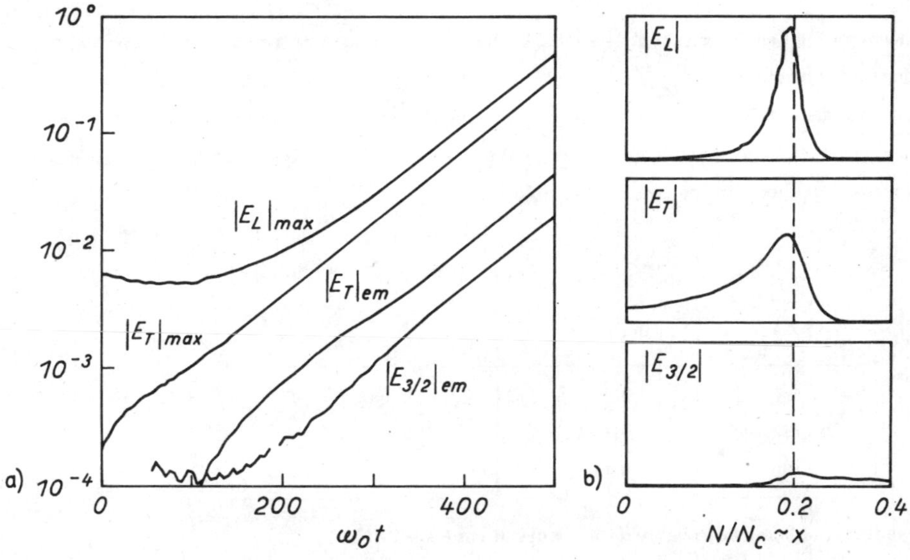

Fig. 7.9 Temporal growth of the maximum amplitudes of the decay waves as well as the vacuum values of the scattered and the $3\omega_o/2$-wave at SRS (a) and spatial amplitude profiles (b). In the latter graphs the horizonzal axis represents the distance along the linearly growing density profile; $v_E/c = 0.1$, $L = 50\lambda_o$, $v_T/c = 0.05$

demonstrated in Fig. 7.9 (b) where spatial profiles of the decay wave amplitudes and the three-halves-wave are plotted. The spatial region around the maximum of the longitudinal decay wave can be considered as the operation regime of the SRS instability. Both forward- and backscattering contribute to the emitted $\omega_o/2$-signal since a forward-propagating decay wave will be totally reflected at its critical density near $N_c/4$. Fig. 7.9 (b) indicates that the three-halves wave will be generated only in forward direction (parallel to the pump wave direction). It may appear, however, also in the backscattered radiation due to total reflection at the layer $N \simeq 9N_c/4$ in an increasing density profile.

Fig. 7.10 shows growth rate versus pump intensity for several density gradient scale lengths. The thresholds for the absolute SRS instability are relatively high, as expected, and the strong

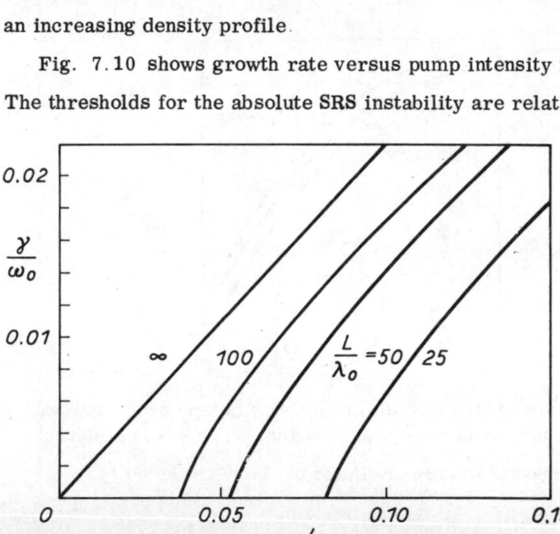

Fig. 7.10 SRS growth rate versus pump amplitude for different density gradient scale lengths ($v_T/c = 0.05$). Curve $L = \infty$ extrapolated

influence of the density gradient is seen. A density scale length of 100 λ_o, for example, induces a threshold of $v_{E,\,thr}/c \simeq 0.038$, corresponding to a pump intensity of $I = 1.8\ 10^{15}$ W/cm^2 for neodymium light. A comparision of the thresholds predicted by Rosenbluth's theory (Eq. (7.19)) and that found numerically by the mode coupling analysis exhibits no significant differences as shown in the small table below.

	$(v_E/c)_{threshold}$ (SRS, $v_T/c = 0.05$)	
L/λ_o	analytical	numerical
25	0.07	0.082
50	0.044	0.053
100	0.028	0.038

These thresholds agree roughly with observed values /13, 25 /.

Fig. 7.11 illustrates the influence of the electron temperature on the SRS instability. The growth rate (a) and the frequency of the backscattered light (b) are plotted as function of the ratio v_T/c for two different density scale lengths at a fixed pump intensity. ($v_E/c = 0.1$). As may be seen from Fig. 7.11 (a) the growth rate does not depend on the electron temperature for $v_T/c < 0.1$, i. e., $T_e < 5$ keV. Only beyond this value the growth rate fairly decreases with increasing electron temperature, obviously as a consequence of enhanced Landau damping. Fig. 7.11 (b) shows the variation of the backscattered frequency with temperature, especially its shift against the nominal line $\omega_o/2$. Due to the relation $\omega_o = \omega_L + \omega_T$ the longitudinal wave and consequently the 3 $\omega_o/2$-wave exhibit just the same shift against $\omega_o/2$ but in the opposite direction. As already predicted by the theory of SRS in a homogeneous plasma, the fre-

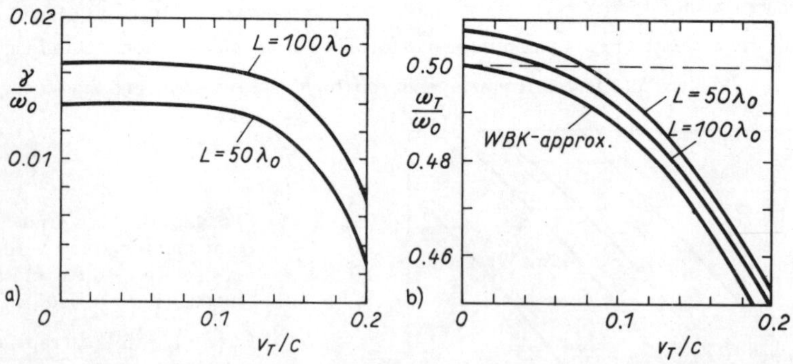

Fig. 7.11 Influence of the electron temperature on SRS: growth rate (a) and backscattered frequency (b) versus v_T/c for two density scale lengths; $v_E/c = 0.1$. Plot (b) shows in addition the scattered frequency predicted by the WKB theory

quency shift relative to $\omega_o/2$ changes its sign with growing electron temperature. While for low temperatures the scattered wave is blue-shifted against $\omega_o/2$, the shift turns to become red for higher temperatures.

Fig. 7.11 (b) shows in addition the dependence of the frequency ω_T of the backscattered wave on the electron temperature as predicted by the WKB theory according to the criterion of perfect matching. Perfect matching means that simultaneously the matching relations for frequency, wave number and the spatial derivative of the wave number are satisfied. Both results are seen to be in good agreement, the better the weaker the density gradient is.

In Fig. 7.12 spatial profiles of the decay wave amplitudes are shown in the case of an extremely high electron temperature ($v_T/c = 0.2$). While the amplitude shape of the scattered wave remains essentially unchanged the longitudinal wave amplitude exhibits now a strong modulation with approximately half the local wavelength. We suggest that this behaviour can be explained in terms of the WKB theory. It predicts a shift of the point x_R of perfect matching from $N_c/4$ toward lower density with increasing temperature. This shift is illustrated in Fig. 7.13 which shows the density at which perfect matching is realized along a linearly increasing profile as function of v_T/c. From this figure one can particularly find that the resonance point x_R is down shifted to a density of $N/N_c \simeq 0.175$ when v_T/c rises to 0.2. Since resonance point and critical layer are then well separated, one has to expect, that the right-propagating longitudinal decay wave that is generated and amplified by the instability near x_R interferes with its oppositely propagating counterpart, produced by total reflection at its critical point. Fig. 7.12 clearly indicates this situation.

Obviously, the shift of x_R is also responsible for the significant drop of the growth rate with increasing temperature (shown in Fig. 7.11 (a)), because the operation region of the instability moves with growing temperature into density regions where the longitudinal wave more and more suffers from enhanced Landau damping. Pump reflection can be included in the model by a continuation of the increasing density profile up to the critical density associated with the pump wave. In order to prevent the spatial computational interval from extending to uninteresting regions where only the pump wave is present, we simulate the high-density part of the profile by a jump of the density from a value near $N_c/4$ to an overcritical one.

Calculations with totally reflected pump wave yield a significant increase in the growth rates

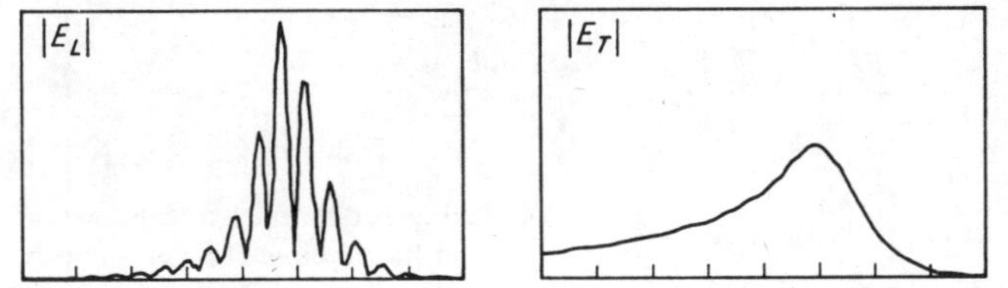

Fig. 7.12 Spatial SRS decay wave amplitude profiles in the case of extremely high electron temperature ($v_T/c = 0.2$, $L = 40\ \lambda_o$)

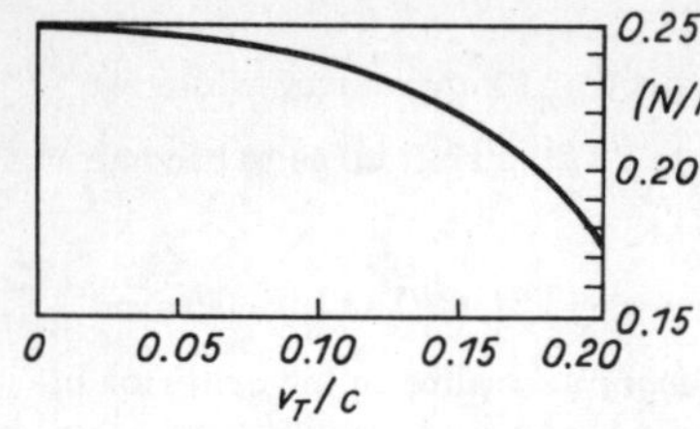

Fig. 7.13 Point $(N/N_c)_{res.}$ of perfect matching at a linear density profile as function of the electron temperature

and a corresponding drop in the thresholds. The absolute SRS threshold induced by a density scale length $L = 100\ \lambda_o$, for example, reduces from $v_E/c = 0.038$ to $v_E/c = 0.02$, corresponding to an intensity of $I = 6 \cdot 10^{14}\ W/cm^2$ for Nd-light. The enhanced growth rate can be understood as a consequence of an enlarged effective length for the three waves to interact, created by the existence of two counterpropagating pump waves The situation is illustrated in a simplified model of a two-step density profile in Fig. 7.14. In case of a single inward propagating pump wave the scattered wave is preferentially generated in back direction and is amplified on passing the low-density part of the profile. When a totally reflected pump wave is present, however, also a forward propagating scattered wave is generated which can further be amplified after turning at the critical density (simulated by the density jump) by interacting with the ingoing pump wave. This results in a doubling of the effective interaction length.

Fig. 7.15 displays spatial amplitude structures of the decay waves when total pump wave reflection occurs. While the transverse amplitude profile has the same shape as in absence of a reflected pump, the plasma wave amplitude is periodically modulated with half the pump wave length, i. e., the standing wave structure of the pump wave is reproduced to a certain extent. This behaviour might be explained along the following line: The transverse decay wave can be considered as to exhibit the weakest spatial variation of the three waves across the operation region of the instability In WKB-terms, we have approximately $k_T \simeq 0$. Thus we may neglect the second spatial derivative in Eq. (7.27) and, assuming steady state conditions, find that E_T is approximately given by

$$E_T \simeq -\frac{e}{2m_e}\ \frac{E_o\ dE_L/dx}{\omega_T\,\omega_o\,\mathcal{E}(\omega_T,\ N)}\ .$$

Introducing this expression into Eq. (7.26) we get the following differential equation for E_L:

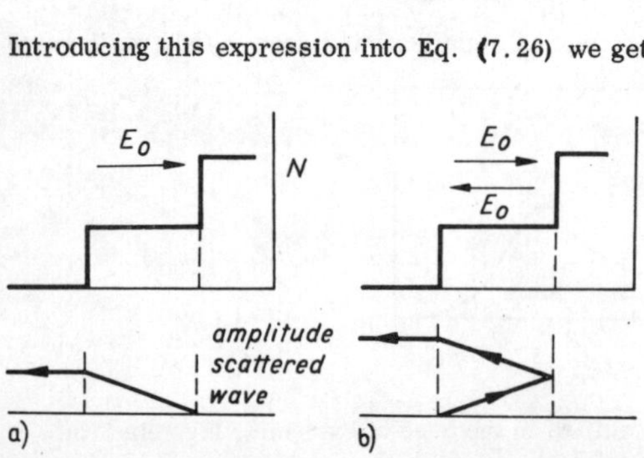

Fig. 7.14 Simplified model to demonstrate the increase of the SRS growth rate when a reflected pump wave is present

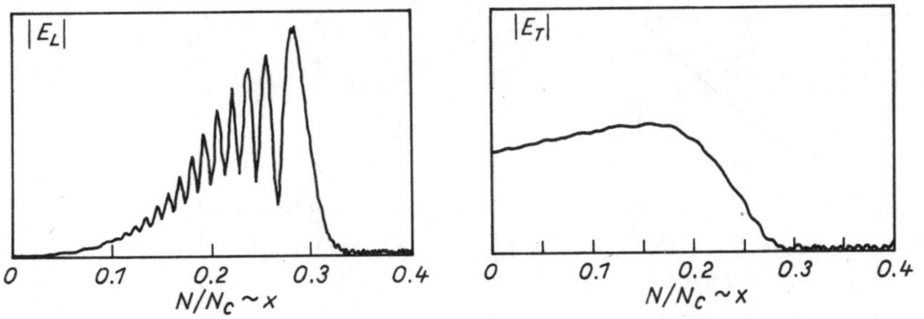

Fig. 7.15 Spatial amplitude profiles of SRS decay waves at the presence of a totally reflected pump wave ($v_T/c = 0.05$, $L = 25\,\lambda_o$)

$$\left\{ 3v_T^2 + \left(\frac{e}{2m_e}\right)^2 \frac{N}{N_c} \frac{|E_o|^2}{\omega_L\,\omega_T\,\mathcal{E}(\omega_T, N)} \right\} \frac{d^2E_L}{dx^2}$$

$$+ \frac{1}{2}\left(\frac{e}{2m_e}\right)^2 \frac{N}{N_c} \frac{d/dx|E_o|^2}{\omega_L\,\omega_T\,\mathcal{E}(\omega_T, N)} \frac{dE_L}{dx} + \omega_L^2\,\mathcal{E}(\omega_L, N)\,E_L = 0 .$$

The two terms containing $|E_o|^2$ exhibit a spatial periodicity with $\lambda_o/2$. Near the turning point of the transverse wave, i. e., for $\mathcal{E}(\omega_T, N) \simeq 0$, they can dominate the equation and force the longitudinal amplitude $|E_L|$ to oscillate in space with $\lambda_o/2$ as shown in the figure.

Two-plasmon decay: As already pointed out within the theory of TPD in a homogeneous plasma, finite propagation angles of the decay waves relative to the pump wave direction are necessary to produce a non-zero growth rate. Thus, we need to complete the set of parameters v_E/c, L, v_T/c by finite k_z-values ($k_{2z} = -k_{1z}$) of the two p-polarized decay waves. For computational reasons, the wave equations for the two associated $3\,\omega_o/2$-waves have not been solved simultaneously.

Fig 7.16 shows analogously to Fig. 7.9 the temporal growth of the maximum decay amplitudes (a) and spatial amplitude profiles of the two decay waves for a typical parameter set ($v_E/c = 0.05$, $L = 100\,\lambda_o$, $v_T/c = 0.05$, $k_z/k_o = 1.0$, k_o free space wave length of the pump wave). There is again an exponential growth in time, indicating the absolute nature of the TPD instability. The spatial amplitude structure confirmes the predictions of the homogeneous theory showing an operation regime of the instability confined to the vicinity of the quarter critical density. Note that the decay waves attain its maximum amplitudes at slightly different positions indicating a small relative frequency shift between both waves. This is in accordance with the homogeneous theory which predict zero growth rate for equal frequencies. The frequency difference is transferred to the three-halves waves, thus we have to expect two slightly shifted $3\,\omega_o/2$-lines originating from TPD, in contrast to the single line produced by SRS. Line splitting of $3\,\omega_o/2$-light has been observed in a number of experiments /8, 14, 16, 19 - 22, 83, 84/.

Fig. 7.17 illustrates the variation of the TPD growth rate with k_z for different density gradient scale lengths (fixed pump intensity $v_E/c = 0.1$).

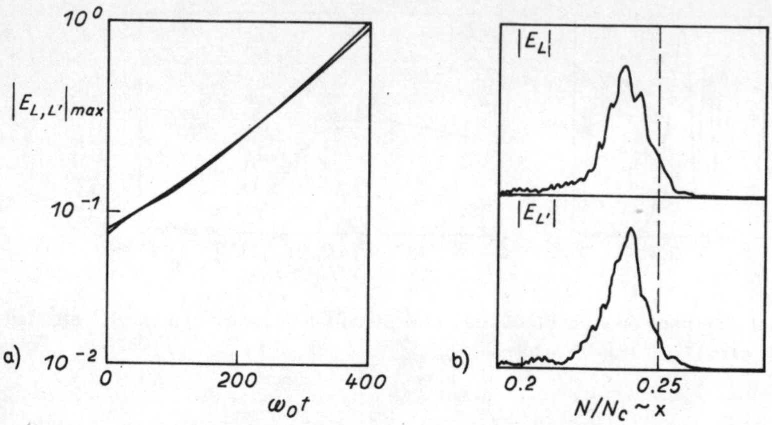

Fig. 7.16 Temporal growth of the maximum amplitudes of the decay waves (a) and spatial amplitude profiles (b) for TPD in an inhomogeneous plasma with a linear density profile; $v_E/c = 0.05$, $L = 100\,\lambda_o$, $v_T/c = 0.05$, $k_z/k_o = 1$

The fact that the curves join the ordinate axis at finite values might appear to be a surprising result because one should expect that the growth rate approaches zero when k_z vanishes. The reason for this behaviour is the following: The integration of the TPD mode coupling system has always been started without assuming a particular polarization of the decay waves; that is, both components E_x, E_z have been given non-zero initial values. This automatically leads to a non-zero growth rate in the case $k_z = 0$ which coincides with the corresponding SRS growth rate. The expected drop of the TPD growth rate to zero for $k_z \rightarrow 0$ occurs when E_z is set equal to zero at the beginning.

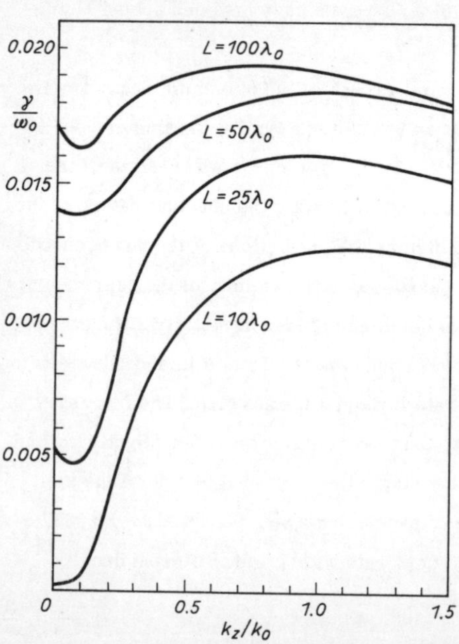

Fig. 7.17 TPD growth rate versus ratio k_z/k_o; k_z wave number of the decay wave perpendicular to the pump wave direction, k_o free space pump wave number; $v_E/c = 0.1$

The graphs enable us to compare TPD and SRS growth rates. First, the predicted different sensitivity to the density gradient of both instabilities is obvious. The SRS growth rate rapidly decreases with increasing density gradient whereas the TPD growth rate is only little affected by a change in the density gradient. At weak gradients ($L > 100\ \lambda_o$) the optimum growth rates of two-plasmon decay (with respect to variation of k_z) are roughly the same as that for SRS. This is in accordance with the homogeneous theory which predicts approximately the same optimum growth rates for both types of instability (cf. Fig. 7.3 and Fig. 7.8). Similar results have been found for lower electron temperatures. From Fig. 7.17 it is seen that TPD operates most effectively in the range $k_z/k_o = 1$. There are observations which indicate optimum TPD growth rates below the value $k_z/k_o = 1$ /46/ as well as above /18, 66/.

The small table

		(v_E/c) threshold (TPD, optimum value of k_z)	
v_T/c	L/λ_o	analytical	numerical
0.05	50	0.01	0.02
0.05	100	0.007	0.01
0.025	100	0.0035	0.007

compares TPD thresholds predicted by the analytical theory (Eq. (7.20)) with those found in mode coupling calculation for three combinations of parameters. The calculations confirm that TPD exhibits significantly lower thresholds than SRS. There are, however, differences of about a factor 2 when compared with the analytical values. A comparision with observed TPD thresholds does not allow a final judgement.

An integration of the nonlinear mode coupling systems including the equation for the pump wave leads to a saturation of both instabilities. The maxima of the decay wave amplitudes, however, saturate at unacceptable high levels. This confirms that pump depletion provides not a effective saturation mechanism of stimulated Raman scattering and two-plasmon decay at high pump amplitudes.

7.6. Odd-integer half-harmonic radiation

As already pointed out at the beginning of this chapter, the emission of half-integer harmonics of the incident radiation and the two decay instabilities SRS, TPD near the quarter critical density are closely related. These harmonics have variously been observed in laser-plasma experiments /1 - 24, 83, 84/. In an inhomogeneous plasma the two-plasmon decay is expected to be the dominant mechanism because of its lower threshold. Typical feature of observed half-harmonic as well as three-halves harmonic spectra is a double-peaked structure with a red-shifted and a blue-shifted wing with respect to the nominal lines $\omega_o/2$, $3\omega_o/2$. These spectra have been observed and frequently reported in the literature /8,11,14,16,19-22,58,66,79-85/. Various theoretical predictions for the spectral structure and the emission intensities of the half-integer harmonic emission have been made though none has yet been established definitely. In most of them the maxima of the two peaks in the spectrum are attributed to the frequencies of the two Langmuir waves comprising the most unstable modes of the two-plasmon decay instability. Since theory predicts the frequency splitting to depend linearly on the local electron temperatur at $N_c/4$, the spectral analysis is of interest mainly for diagnostic purposes. The spectral splitting, however, depends also on other factors geometrical in character, such as angle of incidence and angle of observation which are usually not very well defined in experiment. This reduces the usefulness of the half-integer harmonic emission as a plasma diagnostic, in particular that of the $3\omega_o/2$ spectra.

The generation of odd-integer half-harmonic radiation from laser plasmas can occur via several processes. Common to all of them is the nonlinear generation of Langmuir waves via the primary instabilities near $N_c/4$, preferentially the two-plasmon decay.

The two plasmons generated in the TPD instability exhibit a frequency shift from $\omega_o/2$ which can easily be obtained from the dispersion relation of the two waves and the decay matching conditions to be

$$\frac{\Delta\omega}{\omega_o} = \frac{3}{2}\,\frac{v_T^2 k_o^2}{\omega_o^2}\left(1 - \frac{2\vec{k}_o\vec{k}_L}{k_o^2}\right) \qquad (7.36)$$

($\omega_{L,L'} = \omega_o/2 \pm \Delta\omega$). If their wave vectors are directed close to the gradient direction, these plasmons may undergo mode conversion (inverse resonance absorption) to produce directly electromagnetic waves symmetrically shifted to either side of $\omega_o/2$. Alternatively, the incident light may couple with the plasma waves at $N_c/4$ to generate electromagnetic radiation with frequencies $\omega_H = \omega_o - \omega_{L,L'} = \omega_o/2 \pm \Delta\omega$ (Raman down conversion). Further, emission around the $3\omega_o/2$ harmonic frequency can be generated via coalescence of three plasmons or via Raman up scattering ($\omega_{3H} = \omega_o + \omega_{L,L'}$). In the latter case, the blue wing is produced in forward scatter and the red wing in backscatter. The observation of the blue-shifted frequencies in the backscatter direction can be explained by the reflection of forward-propagating $3\omega_o/2$ radiation at its own critical density $N = 9N_c/4$.

An alternative possibility to explain the spectral structure of the $\omega_o/2$ signal has been pointed out (/75, 81/) where strong self-generated magnetic fields in the plasma are suggested to be responsible for the splitting in frequency in the SRS emission spectrum.

Eq. (7.36) shows that the frequency shift depends both on the local electron temperature and the $\vec{k}$-vector of the decay plasmon. Thus the spectral shifts of the scattered light at $\omega_o/2$, $3\omega_o/2$ contain an unknown parameter represented by the wave vector of the plasmon participating in the generation of this light. It is obvious that the most amplified mode is representative for the choice of k_L. In Fig. 7.7 it is demonstrated, that the k-vector of this mode varies with density and temperature.

For Raman down/up conversion to operate efficiently wavenumber matching conditions are required to be fulfilled in addition to the k-matching conditions of the primary TPD instability. If the harmonic emission is assumed to be generated at the same density as the two-plasmon decay occurs (i. e., if the frequency- and wave vector matching conditions for both processes are imposed for a constant density) then there is a unique solution for the frequency of the harmonic light at a given observation angle resulting in either a red or a blue shift peak in disagreement with the experimental results. Allowing the decay and the scattering to occur at different densities gives a spectrum of emission at all angles because Langmuir waves generated at one point can scatter light at another. This is because the wave number of a propagating plasmon is a sensitive function of the local density in a non-uniform plasma due to the Bohm-Gross dispersion relation.

Moreover, a plasmon excited in the TPD instability may be reflected from higher density layers and the same may happen to the incident wave. Both effects result in a number of different coupling configurations. The situation is further complicated because modes from the broad k-spectrum of the TPD instability may be selected by the combined matching conditions for decay and scattering, that are not identical to the fastest growing modes. The multiplicity of all of these effects makes the analysis of the half-harmonic spectra very difficult. Finally it should be noted that different plasma conditions, different wavelengths of the incident radiation, collisional losses of the scattered waves and various two-dimensional effects may strongly affect the spectra and the emission intensities of the odd-integer half-harmonic radiation.

References

/1/ Bobin, J.L., Decroisette, M., Meyer, B., Vitel, Y.; Phys. Rev. Lett. 30, 594(1973).

/2/ Lee, P., Giovanelli, D.V., Godwin, R.P., McCall, G.H.; Appl. Phys. Lett. 24, 406(1974).

/3/ Jackel, S., Albritton, J., Goldman, E.; Phys. Rev. Lett. 35, 514(1975).

/4/ Pant, H.C., Eidmann, K., Sachsenmeier, R., Sigel, R.; Opt. Comm. 16, 396(1976).

/5/ Jackel, S., Perry, B., Lubin, M.; Phys. Rev. Lett. 37, 95(1976).

/6/ Burnett, N.H., Baldis, H.A., Enright, G.D., Richardson, M.C., Corkum, P.B.; J. Appl. Phys. 48, 3727(1977).

/7/ Bychenko, V.Yu., Zukhorenko, Yu.A., Krokhin, O.N., Rapsov, A.A., Silin, V.P., Sklizkov, G.V., Starodub, A.N., Tikhonchuk, V.T., Shikanov, A.S.; Zh. Eksp. Teor. Fiz. Pisma 26, 500(1977).

/8/ Avrov, A.I., Bychenko, V.Yu., Krokhin, O.N., Pustovalov, V.V., Rupasov, A.A., Silin, V.P., Sklizkov, G.V., Tikhonchuk, V.T., Shikanov, A.S.; Zh. Eksp. Teor. Fiz. 72, 970(1977).

/9/ Aleksandrov, V.V., Vikharev, V.D., Zotov, V.P., Kovalski, N.G., Pergament, M.I.; Zh. Eksp. Teor. Fiz. 24, 10(1978).

/10/ Leonard, T.A., Cover, R.A.; J. Appl. Phys. 50, 3241(1979).

/11/ Carter, P.D., Sim, S.M.L., Barr, H.C., Evans, R.G.; Phys. Rev. Lett. 44, 1407(1980).

/12/ Kilpio, A.W., Maljutin, A.A., Paschinin, P.P.; Pisma Zh. Eksp. Teor. Fiz. 32, 520(1980).

/13/ Elazar, J., Toner, W.T., Wooding, E.R.; Plasma Phys. 23, 813(1981).

/14/ McGoldrick, E., Sim, S.M.L.; Opt. Comm. 40, 433(1982).

/15/ Ebrahim, N.A., Baldis, H.A., Joshi, C., Benesch, R.; Phys. Rev. Lett. 45, 1197(1980).

/16/ Phillion, D.W., Banner, D.L., Campbell, E.M., Turner, R.E., Estabrook, K.G.; Phys. Fluids 25, 1434(1982).

/17/ Turner, R.E., Phillion, D.W., Campbell, E.M., Estabrook, K.G.; Phys. Fluids 26, 579(1983).

/18/ Baldis, H.A., Walsh, C.; Phys. Fluids 26, 1364(1983).

/19/ Powers, L.V., Schroeder, R.J.; Phys. Rev. A29, 2298(1984).

/20/ McGoldrick, E., Sim, S.M.L., Turner, R.E., Willi, O.; Opt. Comm. 50, 107(1984).

/21/ Figueroa, H., Joshi, C., Azechi, H., Ebrahim, N.A., Estabrook, K.; Phys. Fluids 27, 1887(1984).

/22/ Short, R.W., Seka, W., Tanaka, K., Williams, E.A.; Phys. Rev. Lett. 52, 1496(1984).

/23/ Drake, R.P., Turner, R.E., Lashinski, B.F., Estabrook, K.G., Campbell, E.M., Wang, C.L., Phillion, D.W., Williams, E.A., Kruer, W.L.; Phys. Rev. Lett. 53, 1739(1984).

/24/ Seka, W., Williams, E.A., Craxton, R.S., Goldman, L.M., Short, R.W., Tanaka, K.; Phys. Fluids 27, 2181(1984).

/25/ Meyer, J., Bernard, J.E., Hilko, B., Houtman, H., McIntosh, G., Popil, R.; Phys. Rev. A29, 1375(1984).

/26/ Oraevski, V.N., Sagdeev, R.Z.; Zh. Tekhn. Fiz. 32, 1291(1962).

/27/ Silin, V.P.; Zh. Eksp. Teor. Fiz. 47, 2254(1964).
Zh. Eksp. Teor. Fiz. 48, 1679(1965).

/28/ Goldman, M.V.; Ann. Phys. (N.Y.) 38, 95(1966).
Ann. Phys. (N.Y.) 38, 117(1966).

/29/ DuBois, D.F., Goldman, M.V.; Phys. Rev. Lett. 14, 544(1965).
Phys. Rev. 164, 207(1967).

/30/ Sagdeev, R.Z., Galeev, A.A.; Nonlinear Plasma Theory, Benjamin, New York (1969).

/31/ Galeev, A.A., Sagdeev, R.Z.; Nucl. Fusion 13, 603(1973).

/32/ Nishikawa, K.; J. Phys. Soc. Jap. 24, 1152(1968) 916

/33/ Jackson, E.A.; Phys. Rev. 153, 235(1967).

/34/ Kruer, W.L., Dawson, J.M.; Phys. Fluids 14, 1003(1971).

/35/ Forslund, D.W., Kindel, J.M., Lindman, E.L.; Phys. Rev. Lett. 30, 739(1973).

/36/ Drake, J., Kaw, P., Lee, Y.C., Schmidt, G., Liu, C.S., Rosenbluth, M.N.; Phys. Fluids 17, 778(1974).

/37/ Forslund, D.W., Kindel, J.M., Lindman, E.L.; Phys. Fluids 18, 1002(1975).

/38/ Bornatici, M.; J. Plasma Phys. 14, 105(1975).

/39/ Koch, P., Albritton, J.; Phys. Rev. Lett. 34, 1616(1975).

/40/ Klein, H.H., Ott, E., Manheimer, W.M.; Phys. Fluids 18, 1031(1975).

/41/ Estabrook, K., Kruer, W.L.; Phys. Fluids 26, 1892(1983).

/42/ Hiob, E., Barnard, A.J.; Phys. Fluids 26, 3119(1983).

/43/ Powers, L.V., Berger, R.L.; Phys. Fluids 27, 242(1984).

/44/ Estabrook, K., Kruer, W.L.; Phys. Rev. Lett. 53, 465(1984).

/45/ McKinstrie, C.J., Simon, A., Williams, E.A.; Phys. Fluids 27, 2738(1984).

/46/ Villeneuve, D.M., Keck, R.L., Afeyan, B.B., Seka, W., Williams, E.A.; Phys. Fluids 27, 721(1984).

/47/ Meyer, J., Bernard, J.E., Hilko, B., Houtman, H., McIntosh, G., Popil, R.; Phys. Fluids 26, 3162(1983).

/48/ Berger, R.G., Brooks, R.O., Pietrzyk, Z.A.; Phys. Fluids 26, 354(1983).

/49/ Turner, R.E., Estabrook, K., Kaufman, R.L., Bach, D.R., Drake, R.P., Phillion, D.W., Lasinski, B.F., Campbell, E.M., Kruer, W.L., Williams, E.A.; Phys. Rev. Lett. 54, 189(1985).

/50/ Klein, H.H., Manheimer, W.M., Ott, E.; Phys. Rev. Lett. 31, 1187(1973).

/51/ Karttunen, S.J.; Phys. Rev. A23, 2006(1981)

/52/ Rosenbluth, M.N.; Phys. Rev. Lett. 29, 565(1972).

/53/ Liu, C.S., Rosenbluth, M.N., White, R.B.; Phys. Rev. Lett. 31, (1973)697

/54/ Drake, J.F., Lee, Y.C.; Phys. Rev. Lett. 31, 1197(1973).

/55/ Liu, C.S., Rosenbluth, M.N., White, R.B.; Phys. Fluids 17, 1211(1974).

/56/ White, R., Kaw, P., Pesme, D., Rosenbluth, M.N., Laval, G., Huff, R., Varma, R.; Nucl. Fusion 14, 45(1974).

/57/ Lee, Y.C., Kaw, P.K.; Phys. Rev. Lett. 32, 135(1974).

/58/ Liu, C.S., Rosenbluth, M.N.; Phys. Fluids 19, 967(1976).

/59/ Schuss, J.J.; Phys. Fluids 20, 1121(1977).

/60/ Gorbunov, L.M., Domrin, V.I., Ramazashvili, R.R.; Zh. Eksp. Teor. Fiz. 70, 2161(1976).

/61/ Cairns, R.A.; J. Plasma Phys. 12, 169(1974); J. Plasma Phys. 14, 327(1975).

/62/ Mostrom, M.A., Kaufman, A.N.; Phys. Rev. Lett. 42, 644(1979).

/63/ Kruer, W.L., Estabrook, K., Lasinski, B.F., Langdon, A.B.; Phys. Fluids 23, 1326(1980).

/64/ Gorbunov, L.M., Shirokov, A.S.; Fizika Plazmy 6, 663(1980).

/65/ Baumgärtel, K., Sauer, K., Sünder, D.; VII. School Workshop on Plasma Physics, Telawi, USSR (1980).

/66/ Simon, A., Short, R.W., Williams, E.A., Dewardre, T.; Phys. Fluids 26, 3107(1983).

/67/ Koch, P., Williams, E.A.; Phys. Fluids 27, 2346(1984).

/68/ Simon, A., Short, R.W.; Phys. Rev. Lett. 53, 1912(1984).

/69/ Bychenko, W.Yu., Zozulia, A.A., Silin, V.P., Tikhonchuk, W.T.; Preprint 136, FIAN, Moscow (1981).

/70/ Biskamp, D., Welter, H.; Phys. Rev. Lett. 34, 312(1975).

/71/ Forslund, D.W., Kindel, J.M., Lindman, E.L.; Phys. Fluids 18, 1017(1975).

/72/ Langdon, A.B., Lasinski, B.F., Kruer, W.L.; Phys. Rev. Lett. 43, 133(1979).

/73/ Estabrook, K.G., Kruer, W.L., Lasinski, B.F.; Phys. Rev. Lett. 45, 1399(1980).

/74/ Kruer, W.L., Estabrook, K.G., Langdon, A.B.; Phys. Fluids 23, 1232(1980).

/75/ Barr, H.C., Boyd, T.M.J., Gardner, G.A., Rankin, R.; Phys. Rev. Lett. 53, 462(1984).

/76/ Fedorchenko, A.M., Kozarenko, N.J.; Absolute and convective instability in plasmas and solids (in Russian), Nauka, Moscow (1981).

/77/ Bers, A.; in: Handbook of Plasma Physics. Eds. M.N. Rosenbluth, R.Z. Sagdeev Vol. I, p. 451. North-Holland Publ. Comp., 1983.

/78/ Gorbunov, L.M.; Zh. Eksp. Teor. Fiz. 67, 1386(1974).

/79/ Gusakov, E.Z.; Pisma Zh. Eksp. Teor. Fiz. 3, 1219(1977).

/80/ Bychenkov, V.Yu., Zozulja, A.A., Silin, V.P., Tikhonchuk, V.T.; Beitr. Plasmaphys. 23, 331(1983).

/81/ Barr, H.C., Boyd, T.J.M., Gardner, L.R.T., Rankin, R.; Phys. Fluids 27, 2730(1984).

/82/ Powers, L.V., Berger, R.L.; Phys. Fluids 28, 2419(1985).

/83/ Aboites, V., Hughes, T.P., McGoldrick, E., Sim, S.M.L., Karttunen, S.J., Evans, R.G.; Phys. Fluids 28, 2555(1985).

/84/ Seka, W., Afeyan, B.B., Boni, R., Goldman, L.M., Short, R.W., Tanaka, K., Johnston, T.W.; Phys. Fluids 28, 2570(1985).

/85/ Berger, R.L., Powers, L.V.; Phys. Fluids 28, 2895(1985).

Part IV Low-frequency nonlinear processes

The preceding chapters dealt with high-frequency phenomena such that ions could be taken as fixed. Now we consider interaction processes taking place on a much slower time scale, allowing the ions to react on the high-frequency field via the ponderomotive force. The corresponding mathematical framework is given by the system of coupled partial differential Eqs. (3.12) - (3.14) derived in Chapter 3 which governs the low-frequency response of a plasma to a high-frequency wave. Hence, we restrict ourselves to problems in which the fundamental physical source of nonlinearity can be traced back to the modification of the local value of the plasma density due to the ponderomotive force.

Neglecting the interaction with a high-frequency field, a fully ionized plasma without magnetic field exhibits only ion acoustic waves as interesting low-frequency phenomenon in the case $T_e \gg T_i$. In the presence of weak nonlinearities and weak dispersion these waves are governed by the Korteweg-de Vries equation /1/.

Taking into account the interaction with a high-frequency field leads to a rich variety of nonlinear phenomena such as Brillouin scattering, anomalous skin effect, soliton formation, wave collaps, superreflection etc. The key concept for understanding all these processes is the plasma dynamics under the action of the ponderomotive force of a strong high-frequency field.

The theory of this field was pioneered by Gorbunov who gave also a first review /2/. A simplified approach is possible if one restricts to problems in which the field-induced density disturbance remains small compared to the equilibrium plasma density. Then the equations for the plasma dynamics can be linearized and the basic system (3.14) - (3.16) reduces to two coupled wave equations for low-frequency disturbances and the slowly varying envelope of the high-frequency field. In a uniform plasma an additional simplification arises owing to the decoupling of longitudinal and transverse high-frequency fields.

The interaction of a transverse high-frequency wave with ion acoustic waves in a subcritical plasma is known as stimulated Brillouin scattering. Interaction of transverse waves with slightly overcritical plasmas has been studied in terms of envelope-solitons in /3/.

The interaction of longitudinal high-frequency waves with ion acoustic waves was first studied by Zakharov /4/. The corresponding basic equations (Zakharov system) and methods of their solution have recently attracted much interest and a lot of papers have appeared in this context (see, for example /5 - 15/). They are partly related to problems of plasma wave turbulence

(Langmuir turbulence), excited by an external pump, as an electron beam or a laser field. As elements of strong turbulence, envelope solitons become significant in the one-dimensional case. In two or three dimensions, however, these solutions are not stabil and tend to collapse.

In the quasi-stationary limit, when the second order time derivative in the ion acoustic wave equation is negligible, the Zakharov system turns to the non-linear Schrödinger equation, the solutions of which have also been intensively studied /16 - 19/.

Concerning strongly nonlinear processes, a more rigorous approach is necessary, which starts from the fully nonlinear equations of the plasma dynamics. First effort was given to equilibrium configurations between plasma and field /20/. Non-equilibrium processes can easily be treated, as a first step, in the stationary approximation. The resulting ordinary differential equations can be solved analytically to a certain extent /21/. Mathematical modelling of strong, nonstationary interaction between a high-frequency field and the plasma, however, is only possible on the basis of numerical solutions of the basic equations.

We concentrate here on stimulated Mandelstam-Brillouin scattering (Chapter 9), nonlinear stationary waves (Chapter 10) and several examples of strongly nonlinear instationary interaction (Chapter 11). As preliminary considerations, Section 8.1 presents some relevant properties of ion acoustic wave propagation and Section 8.2 gives an introduction to the Zakharov system, the nonlinear Schrödinger equation and related problems. The well-developed theories of the Korteweg-de Vries equation, the Zakharov system and the nonlinear Schrödinger equation shall not be presented here.

For details concerning strong Langmuir turbulence, collapse etc. we refer to the review paper of Goldman /22/ and the bibliography contained in it. A survey of the theory of general nonlinear wave equations including the Korteweg-de Vries equation, the Zakharov system and the nonlinear Schrödinger equation with emphasis given on mathematical rigour can be found in the book of Dodd, Eilbeck, Gibbon and Morris /31/.

8 Linearized plasma dynamics

8.1. Ion acoustic waves

In this section we shall briefly outline some properties of ion acoustic wave propagation. We shall start with Eqs. (3.14), (3.15), ignoring at the moment the external electric field. The two equations

$$\frac{\partial N}{\partial t} + \operatorname{div}(N \vec{v}) = 0 , \tag{8.1}$$

$$\frac{\partial v}{\partial t} + (\vec{v}\ \mathrm{grad})\ \vec{v} = - c_s^2 \frac{1}{N} \operatorname{grad} N \tag{8.2}$$

describe the dynamics of an isothermal compressible fluid. They differ from that of ordinary gas dynamics only in the expression (8.4) for the ion acoustic velocity c_s. In case of small deviations δN from an equilibrium density N the equations can be linearized and combined to give the wave equation

$$\frac{\partial^2 \delta N}{\partial t^2} - c_s^2 \triangle \delta N = 0 \tag{8.3}$$

which describes dispersion-free and undamped propagation of ion acoustic waves with the velocity

$$c_s = \left\{ \frac{k_B (\gamma_i T_i + ZTe)}{m_e} \right\}^{1/2} \tag{8.4}$$

From ordinary gas dynamics it is well-known, that a wave equation of type (8.3) with $c_s = (k_B T/m)^{1/2}$ is applicable to describe acoustic waves, as long as its frequency is small against the mean collision frequency and its wave length is large compared to the mean-free path of the particles. Ion acoustic waves, however, represent a completely collision-free phenomenon in plasmas with $T_e \gg T_i$, associated with weak charge separation and thus are entirely different in nature from sound waves in neutral gases /23/.

To illustrate this point, Fig. 8.1 shows dispersion curves of ion acoustic waves as predicted by kinetic theory. The ratio Im(k)/Re(k) and the normalized phase velocity v_{ph}/c_s are plotted versus ω/ω_{pi} for different values of the temperature ratio T_e/T_i. $\omega_{pi} = (m_e/m_i)^{1/2}\ \omega_p$ is

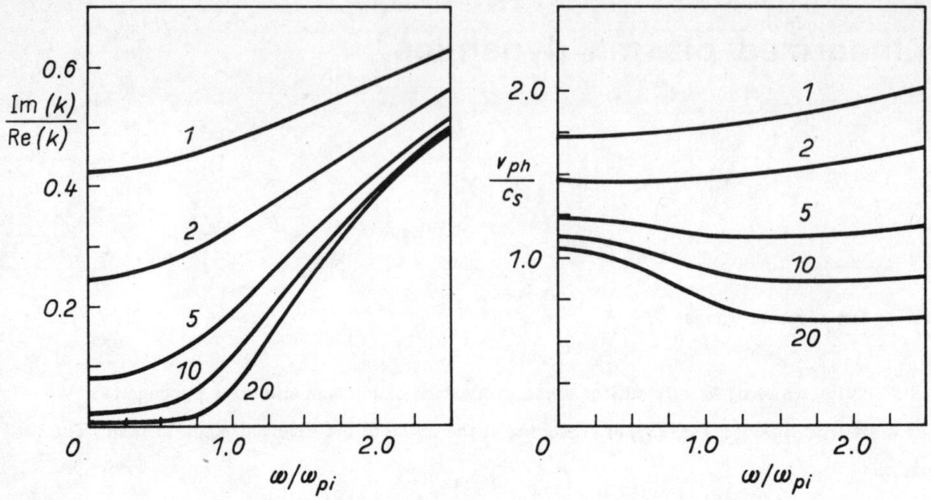

Fig. 8.1 Dispersion of ion acoustic waves predicted by kinetic theory; Re(k) /Im(k) (a) and normalized phase velocity v_{ph}/c_s (b) versus ω/ω_{pi} for different temperature ratios T_e/T_i (curve parameter)

the ion plasma frequency. From Fig. 8.1 it is seen, that weakly damped dispersion-free wave propagation is only possible for $T_e \gg T_i$ in the frequency range below ω_{pi}. Strong ion Landau damping dominates for $T_e \simeq T_i$. For $T_e \gg T_i$ strong damping sets in only for frequencies exceeding the ion plasma frequency. This corresponds to wavelengths being comparable to the Debye length.

Ion Landau damping can be accounted for in the wave Eq. (8.3) by introducing an additional (temperature-dependent) damping term:

$$\frac{\partial^2 \delta N}{\partial t^2} + 2\sigma \frac{\partial \delta N}{\partial t} - c_s^2 \Delta \delta N = 0. \tag{8.5}$$

According to this equation, a spatially homogeneous disturbance relaxes as $\propto \exp(-\sigma t)$. σ is the ion acoustic damping decrement. With a properly chosen σ Eq. (8.5) is capable of describing ion acoustic waves with frequencies lower than the ion plasma frequency and with wavelengths large compared to the Debye length.

Weakly nonlinear ion acoustic waves

The theory of ordinary gas dynamics predicts that the main nonlinear distortion of a large-amplitude monochromatic acoustical wave will be the steepening of the wave fronts. This steepening can be understood in terms of resonant harmonic generation. Obviously, ion acoustic waves behave in quite the same manner as far as they obey Eq. (8.5). We shall now recall some properties of acoustical wave propagation under the influence of a weak nonlinearity.

Using the ansatz for simple waves $N = N(v)$, a modified wave equation containing a quadratic nonlinear term can be derived from Eqs. (8.1), (8.2) by a standard procedure /24/. Considering the one-dimensional version of the basic equations

$$\frac{\partial N}{\partial t} + \frac{\partial}{\partial x}(Nv) = 0, \tag{8.6}$$

$$\frac{\partial v}{\partial t} + v\frac{\partial v}{\partial x} = -c_s^2 \frac{1}{N}\frac{\partial N}{\partial x}, \tag{8.7}$$

introducing the ansatz $N = N(v)$, multiplying Eq. (8.6) by N and Eq. (8.7) by dN/dv and combining both equations yields the relation

$$c_s^2\left(\frac{dN}{dv}\right)^2 = N^2, \quad \text{i. e.,} \quad c_s\frac{dN}{dv} = \pm N. \tag{8.8}$$

This gives in combination with Eq. (8.6) the nonlinear wave equation

$$\frac{\partial v}{\partial t} + (v \pm c_s)\frac{\partial v}{\partial x} = 0 \tag{8.9}$$

for simple waves. Due to $v/c_s \simeq \delta N/N = q$, a similar equation is approximately satisfied also by the density disturbance

$$\frac{\partial q}{\partial t} + c_s(q \pm 1)\frac{\partial q}{\partial x} = 0. \tag{8.10}$$

The different signs correspond to propagation in positive or negative x-direction.

Wave breaking

The solution $q(x, t)$, which develops in time according to Eq. (8.10) from an initial profile $q(x, o) = Q(x)$, can immediately be written as

$$q(x, t) = Q(x - c_s(q + 1)t) \tag{8.11}$$

(positive sign taken in Eq. (8.10)). Thus each part of the density profile $q(x, t)$ propagates with constant velocity, given by $c_s(1 + q)$. Consequently, the initial wave form is distorted and those parts of the profile with decreasing density along the direction of propagation suffer from steepening. The steepening proceeds until the spatial derivative dq/dx becomes infinite. At that moment, t_c, a discontinuity is formed and the profile has almost saw-tooth shape. After that time, the solution remains no longer single-valued and thus becomes physically meaningless. t_c is given by the combined condition /25/

$$\left.\frac{dx}{dq}\right/_{t=t_c} = 0, \quad \left.\frac{d^2x}{dq^2}\right/_{t=t_c} = 0. \tag{8.12}$$

For a sinusoidal initial profile

$$q(x, 0) = q_o \sin kx, \tag{8.13}$$

this gives

$$t_c = \frac{1}{kc_s q_o}. \tag{8.14}$$

The subsequent time evolution is characterized by maintaining the saw-tooth shape with a relaxing amplitude. This stage of the time evolution can easily be treated analytically. Let

$$q(x, t_c) = q_o\left(\frac{kx}{\pi} - 1\right), \qquad 0 \leq x \leq \frac{2\pi}{k} \tag{8.15}$$

be the saw-tooth profile at the time t_c. The solution evolving from this profile according to Eq. (8.10) is

$$q(x, t) = \frac{q_o}{1 + \frac{1}{\pi}\frac{(t - t_c)}{t_c}}\left\{\frac{k(x - c_s t)}{\pi} - 1\right\}. \tag{8.16}$$

This profile relaxes in time as $1/\left\{1 + (t - t_c)/\pi t_c\right\}$, thus showing irreversible behaviour, although Eq. (8.10) does not contain any dissipative term. This irreversible feature is produced via transformation of streaming energy into thermal energy as the fluid passes the discontinuity, similar to the process occurring at a shock front. This point is illustrated in Fig. 8.2. As a result of a numerical integration of Eq. (8.10) it shows the time evolution of a sinusoidal initial profile. The two different stages of the evolution are clearly seen: first the approach toward the saw-tooth profile and subsequently the relaxation without a significant change in profile geometry.

The spatial boundary problem associated with Eq. (8.10) can be treated analogously /24/. Let $q(0, t) = q_o \sin \omega t$ be a periodic disturbance excerted on the plasma at $x = 0$, then the critical distance x_c, where wave breaking occurs, is approximately given by

$$x_c = \frac{c_s}{\omega q_o}. \tag{8.17}$$

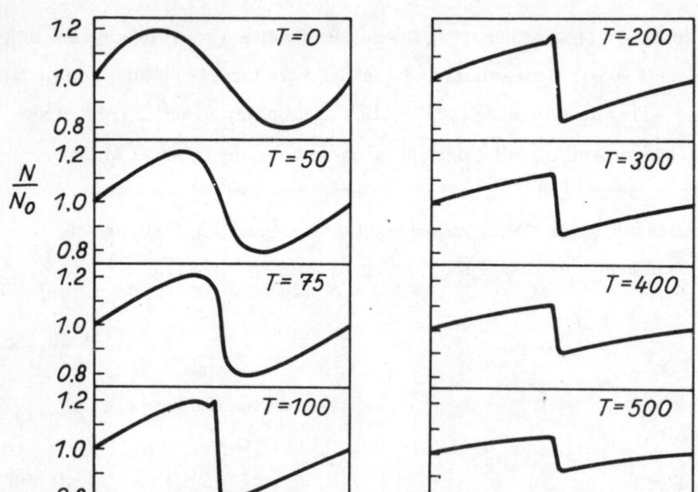

Fig. 8.2 Time evolution of an ion acoustic wave of finite amplitude with sinusoidal initial shape according to Eq. (8.10)

Ion acoustic harmonic generation

An alternative approach to solve Eq. (8.10) in the case of monochromatic excitation is a spectral analysis. It is based on the representation of the density disturbance q (x, t) as superposition of the harmonics of the initial wave with frequency ω and wavenumber k ,

$$q(x, t) = i \sum_{m=-M}^{M} q_m(x, t)\, e^{im(kx - \omega t)} . \tag{8.18}$$

q_m (x, t) are the slowly varying complex amplitudes of the harmonics. Since the dispersion relation is of the form $\omega = c_s k$, the harmonics will also be normal modes, and, being driven resonantly, they will grow in space and time. Thus, a cumulative enrichment of the wave spectrum in higher harmonics will occur. Owing to q being real, each q_m must obey the relation $q_{-m} = q_m^*$. M represents the highest harmonic number included. Introducing the ansatz (8.18) into Eq. (8.10) yields the system

$$\frac{\partial q_n}{\partial t} + c_s \frac{\partial q_n}{\partial x} = \omega n \sum_{m=\left[\frac{n+1}{2}\right]}^{M} c_{m,n}\, q_m q_{n-m} \tag{8.19}$$

where $[z]$ denotes the largest integer less or equal to z and

$$c_{m,n} = \begin{cases} 1 \text{ if } m \neq n/2 \\ \frac{1}{2} \text{ if } m = n/2 \end{cases} .$$

As an example, we write system (8.19) explicitely taking into account 5 harmonics:

$$\left(\frac{\partial}{\partial t} + c_s \frac{\partial}{\partial x}\right) q_1 = \omega (- q_2 q_1^* - q_3 q_2^* - q_4 q_3^* - q_5 q_4^*) ,$$

$$\left(\frac{\partial}{\partial t} + c_s \frac{\partial}{\partial x}\right) q_2 = 2\omega (- q_3 q_1^* - q_4 q_2^* - q_5 q_3^* + \frac{1}{2} q_1^2) ,$$

$$\left(\frac{\partial}{\partial t} + c_s \frac{\partial}{\partial x}\right) q_3 = 3\omega (- q_4 q_1^* - q_5 q_2^* + q_1 q_2) ,$$

$$\left(\frac{\partial}{\partial t} + c_s \frac{\partial}{\partial x}\right) q_4 = 4\omega (- q_5 q_1^* + q_1 q_3 + \frac{1}{2} q_2^2) ,$$

$$\left(\frac{\partial}{\partial t} + c_s \frac{\partial}{\partial x}\right) q_5 = 5\omega (q_1 q_4 + q_2 q_3). \tag{8.20}$$

Cancelling one of the partial derivatives at the left-hand sides, the system can be adapted to describe either a pure time evolution or a boundary value problem. It is obvious from Eqs. (8.20) that real initial values q_i result in q_i being real for $t > 0$. Further, the q_i are easily found to satisfy the conservation law

$$|q_1|^2 + |q_2|^2 + \ldots + |q_M|^2 = \text{const.} \qquad (8.21)$$

Obviously, the spectral analysis also fails in describing the profile development after the occurrence of a discontinuity because Eq. (8.21) is violated during the relaxation phase. Nevertheless, it is apparent that profile steepening is equivalent to the successive generation of higher harmonics.

So far we have ignored dispersion of ion acoustic waves. From the two-fluid plasma equations, however, a modified version of Eq. (8.10) can be derived /26/, if charge separation is taken into account:

$$\frac{\partial q}{\partial t} + c_s q \frac{\partial q}{\partial y} + \frac{\lambda_D^2}{2} \frac{\partial^3 q}{\partial y^3} = 0 \qquad (8.22)$$

($y = x - c_s t$). This equation, well-known as Korteweg-de Vries equation, is capable of describing weakly nonlinear and weakly dispersive ion acoustic waves. It has successfully been applied to various wave phenomena in other fields of physics. Solutions of the KdV-equation has been studied intensively, particularly in terms of solitons /1, 27/.

Dispersion prevents the harmonics of a normal mode to be normal modes themselves. Consequently, the harmonics of a large-amplitude wave usually remain at a low amplitude. As far as the time evolution of a spatially periodic initial profile is concerned, the KdV-equation predicts the generation of a chain of ion acoustic solitons instead of wave breaking /1, 28/.

8.2. Zakharov system, nonlinear Schrödinger equation

We shall continue the considerations of the previous section by including a high-frequency field and looking for equations that describe its interaction with the plasma to lowest order of nonlinearity. As pointed out already, in a homogeneous plasma Eq. (3.16) can be split up into two separate equations for the transverse and the longitudinal component of the high-frequency field. Representing the slowly varying complex amplitude $\vec{E}_\omega$ as $\vec{E}_\omega = \vec{E}_T + \vec{E}_L$ and using the approximation

$$\varepsilon(\omega, N + \delta N) \simeq \varepsilon(\omega, N) - \frac{\omega_p^2}{\omega^2} \frac{\delta N}{N}$$

for the dielectric function we find for the transverse amplitude $\vec{E}_T$ the equation

$$2i\omega \frac{\partial \vec{E}_T}{\partial t} + c^2 \Delta \vec{E}_T + \omega^2 \varepsilon(\omega, N) \vec{E}_T - \omega_p^2 \frac{\delta N}{N} \vec{E}_T = 0 . \qquad (8.23)$$

Linearization of the equations of the plasma dynamics yields, on the other hand, for the density disturbance δN the wave equation

$$\frac{\partial^2 \delta N}{\partial t^2} + 2\sigma \frac{\partial \delta N}{\partial t} - c_s^2 \triangle \delta N = \frac{1}{4} N \frac{e^2}{m_e m_i \omega^2} \triangle |\vec{E}_T|^2 \tag{8.24}$$

where again the phenomenological damping term, introduced in the preceding section, is included. The source term on the right-hand side of Eq. (8.24) represents the ponderomotive force of the high-frequency field.

The two coupled Eqs. (8.23), (8.24) are relevant, e. g., for describing stimulated Brillouin scattering, which occurs, when an electromagnetic wave penetrates an underdense plasma.

The interaction of an electromagnetic wave with a slightly overcritical plasma has first been studied by Karpman /3/. The skin effect, predicted by linear theory, is nonlinearly modified and the field can penetrate the plasma in form of envelope solitions.

Considering the envelope equation for a longitudinal field, the dielectric constant $\mathcal{E}(\omega, N)$ does usually not enter since one can set $\omega = \omega_p$. This is justified because longitudinal disturbances in a plasma are restricted with respect to the frequency to the region $\omega \gtrsim \omega_p$. Deviations from the $\exp(-i\omega_p t)$-dependence will be accounted for by the first-order time derivative. Since $\mathcal{E}(\omega_p, N) = 0$ we have

$$2i\omega_p \frac{\partial \vec{E}_L}{\partial t} + 3v_T^2 \triangle \vec{E}_L - \omega_p^2 \frac{\delta N}{N} \vec{E}_L = 0, \tag{8.25}$$

$$\frac{\partial^2 \delta N}{\partial t^2} + 2\sigma \frac{\partial \delta N}{\partial t} - c_s^2 \triangle \delta N = \frac{1}{4} N \frac{e^2}{m_e m_i \omega^2} \triangle |\vec{E}_L|^2 . \tag{8.26}$$

This system was first studied by Zakharov /4/ and since then it has attracted considerable attention /5 - 15/.

The underlying problem concerns mainly the time evolution of an ensemble of Langmuir waves with wavelengths $\lambda \gg \lambda_D$, excited, for example, by an external source, as an electron beam or a laser field. The general features are as follows: Nonlinear interactions among these waves lead to a concentration of the waves at the long-wavelenght tail of the spectrum. Vedenov and Rudakov /5/ showed that such a Langmuir condensate is instabil with respect to a spatial modulation (modulational instability). The wave ensemble tends to be localized and trapped in a locally underdense plasma region which is created as a result of expulsion of plasma particles from the high-field region by the ponderomotive force. Zakharov /4/ investigated the nonlinear stage of this instability with the help of Eqs. (8.25), (8.26) and showed that in the two-dimensional case the energy concentration merges eventually into a collapse-like behaviour.

In one dimension, the self-modulation of Langmuir waves proceeds differently, leading to soliton formation instead of collapse /6 - 19/. The one-dimensional version of the Zakharov system has been studied extensively by means of both numerical and analytical methods (see, for instance /14/, where one can find a bibliography). The theory is dominated by the soliton concept. Special topics are: formation of solitons from initial conditions, interaction between solitons and of solitons with ion acoustic waves such as soliton scattering accompanied by emission of ion sound, fusion of solitons, fission of solitons accompanied with absorption of ion sound. Langmuir solitons therefore can be created or annihilated by interaction with ion acoustic waves. Selection rules for these processes based on the conservation laws of the Zakharov system are given in /13/.

Numerical calculations on the modulation instability in two dimensions showing instability of solitary solutions and the development of collapse, have been performed in /12, 15/.

In the limit of quasistationary processes, where the ion inertia does not come into play, the second-order time derivative in the ion acoustic wave equation can be neglected and Eq. (8.26) gives the "adiabatic" relation

$$N = -\frac{1}{16\pi k_B T_e} |E_L|^2 . \tag{8.27}$$

Introducing this relation into Eq. (8.25) leads to the nonlinear Schrödinger (NLS) equation

$$2i\omega_p \frac{\partial \vec{E}_L}{\partial t} + 3v_T^2 \Delta \vec{E}_L + \frac{e^2}{4m_e k_B T_e} |E_L|^2 \vec{E}_L = 0 . \tag{8.28}$$

With τ being a characteristic time and l being a characteristic distance of a process, the NLS equation applies if

$$\tau^2 \ll (l/c_s)^2 .$$

With other words, the NLS equation (8.28) is appropriate if the group velocity of the high-frequency field is small compared with the ion acoustic speed. The NLS equation has been solved, for example, by Zakharov and Shabat /16/ with the help of the inverse scattering method.

A two-fold modified version of the NLS equation has been used to model the excitation of Langmuir waves by an external source in an inhomogeneous plasma /17/. Taking into account an inhomogeneity of the plasma density gives rise to a re-appearance of the dielectric function in Eq. (8.25), in analogy to Eq. (8.23). The second modification concerns the inclusion of the external driver. It is assumed to have the form of a spatially homogeneous capacitor field of frequency ω and amplitude E_0. It appears as a source term in the NLS equation and simulates, for example, the externally driven longitudinal field of an obliquely incident p-polarized electromagnetic wave.

Assuming a linear density profile with scale length L and choosing $x = 0$ to be the critical point (i. e., $N/N_c = 1 - x/L$), the dielectric constant can be written as

$$\varepsilon(\omega, x) = -\frac{x}{L} + i\frac{\nu}{\omega} .$$

The modified NLS equation then has the form

$$2i\omega \frac{\partial E_L}{\partial t} + 3v_T^2 \frac{\partial^2 E_L}{\partial x^2} + \omega^2 \left[-\frac{x}{L} + i\frac{\nu}{\omega} \right] E_L + \frac{e^2}{4m_e k_B T_e} \left| E_L \right|^2 E_L = E_o . \tag{8.29}$$

Special attention is given in /18, 19/ to the excitation of solitons by the external field. Eq. (8.29) has further been studied analytically in /29, 30/.

References

/1/ Karpman, V.I.; Nonlinear waves in dispersive media, Pitman, London 1975.

/2/ Gorbunov, L.M.; Usp. Fiz. Nauk 109, 631(1973).

/3/ Karpman, V.I.; Plasma Phys. 13, 477(1971).

/4/ Zakharov, V.E.; Zh. Eksp. Teor. Fiz. 62, 1745(1972).

/5/ Vedenov, A.A., Rudakov, L.I.; Doklady Akad. Nauk SSSR 159, 767(1964).

/6/ Kingsep, A.S., Rudakov, L.I., Sudan, R.N.; Phys. Rev. Lett. 31, 1482(1973).

/7/ Degtyarev, L.M., Makhankov, V.G., Rudakov, L.I.; Zh. Eksp. Teor. Fiz. 67, 533(1974).

/8/ Yajima, N., Oikawa, M.; Progr. Theor. Phys. 56, 1719(1976).

/9/ Karpman, V.I.; Physica Scripta 11, 263(1975).

/10/ Pereira, N.R., Sudan, R.N., Denavit, J.; Phys. Fluids 20, 271(1977).

/11/ Pereira, N.R.; Phys. Fluids 20, 750(1977).

/12/ Pereira, N.R., Sudan, R.N., Denavit, J.; Phys. Fluids 20, 936(1977).

/13/ Gibbons, J., Thornhill, S.G., Wardrop, M.J., ter Haar, D.; J. Plasma Phys. 17, 153(1977).

/14/ Thornhill, S.G., ter Haar, D.; Physics Reports 43, 53(1978).

/15/ Anisimov, S.I., Berezovski, M.A., Ivanov, M.F., Petrov, I.V., Rubenchik, A.M., Zakharov, V.E.; Phys. Lett. 92A, 32(1982).

/16/ Zakharov, V.E., Shabat, A.; Zh. Eksp. Teor. Fiz. 61, 118(1971).

/17/ Morales, G.J., Lee, Y.C.; Phys. Rev. Lett. 33, 1016(1974)
Phys. Fluids 20, 1135(1977).

/18/ Chen, H.H., Liu, C.S.; Phys. Rev. Lett. 39, 1147(1977).

/19/ Anderson, D., Bondeson, A., Lisak, M.; J. Plasma Phys. 22, 339(1979).

/20/ Volkov, T.V.; in: Plasma Physics (ed. Leontovich) Vol. 3, 395(1959).

/21/ Vaskov, V.V., Karpman, V.I.; Zh. Tekhn. Fiz. 41, 676(1971).

/22/ Goldman, M.V.; Rev. Mod. Phys. 56, 709(1984).

/23/ Fried, B.D., Gould, R.W.; Phys. Fluids 4, 139(1961).

/24/ Rudenko, O.V., Soluyan, S.I.; Theoretical foundations of nonlinear acoustics. Consultants Bureau, New York and London, 1977.

/25/ Landau, L.D., Lifshitz, E.M.; Lehrbuch der Theor. Physik, Bd. IV, § 94 Akademieverlag Berlin, 1966.

/26/ Washimi, H., Taniuti, T.; Phys. Rev. Lett. 17, 996(1966).

/27/ Scott, A.C., Chu, F.Y.F., McLaughlin, D.W.; Proc. IEEE 61, 1443(1973).

/28/ Baumgärtel, K., Sauer, K.; in: Physik und Technik des Plasmas. Ed.: Physikalische Gesellschaft der DDR Karl-Marx-Stadt, 1974, p. 121.
II. Intern. Conf. on Plasma Theory, Kiev, 1974, Abstr. Contr. Papers, p. 82.

/29/ Karpman, V.I., Maslov, E.M.; Phys. Lett. 60A, 307(1977).

/30/ Shukla, P.K., Spatschek, K.H.; J. Plasma Phys. 19, 387(1978).

/31/ Dodd, R.K., Eilbeck, J.C., Gibbon, J.D., Morris, H.C.; Solitons and Nonlinear Wave equations, Academic Press, 1982.

9 Stimulated Mandelstam-Brillouin scattering (SBS)

Stimulated Mandelstam-Brillouin scattering (SBS) is another representative of the family of parametric instabilities that can occur in the subcritical region of a plasma irradiated by an electromagnetic wave (see Fig. 7.1). It refers to the parametrically unstable scattering of the incident wave by longitudinal ion acoustic waves; thus, this process may be regarded as the low-frequency counterpart of stimulated Raman scattering. Ordinary Brillouin scattering arises as a result of the nonlinear beating of the incident wave with a proper low-frequency ion acoustic mode such that frequency and wave number matching relations are satisfied. The process develops unstable, when the ponderomotive force produced by the superposition of pump wave and scattered wave is strong enough to drive the original ion acoustic mode. Then a feedback mechanism will be established and scattered wave and ion acoustic wave exponentiate together.

Characteristic features of stimulated Mandelstam-Brillouin scattering are: (1) The frequency shift produced by the scattering is small because an ion acoustic wave with a wavelength comparable to that of an electromagnetic wave with frequency ω in the plasma has a frequency much less than ω. (2) Owing to this fact the SBS instability offers the possibility for almost complete reflection of the incoming radiation (according to the Manley-Rowe relation). (3) Parametric instabilities involving ion acoustic modes exhibit lower thresholds than those involving Langmuir waves; hence SBS is expected to be excited first in experiments. (4) SBS is not restricted to occur only in density regions below quarter critical density but can operate in the whole subcritical region of a plasma.

A simple theoretical access to stimulated Brillouin scattering is provided by the phenomenon of Bragg reflection, well-known from linear optics. While an electromagnetic wave is transmitted by a fairly undercritical plasma of uniform density, the wave can be completely reflected if the density is periodically modulated in space appropriately. Maximum reflectivity is achieved if the period of the density modulation coincides with one-half the wavelength. This effect is demonstrated in Fig. 9.1. It is related to a situation in which an electromagnetic wave penetrates a subcritical plasma whose density varies in space as shown in the left part of the figure. The plasma has an averaged density of $N = 0.3\ N_c$ and is of finite length. The fraction of the reflected energy is plotted versus the ratio l/λ where l is the periodicity length of the density structure and λ is the wave length taken for the averaged density. A significant peak appears at

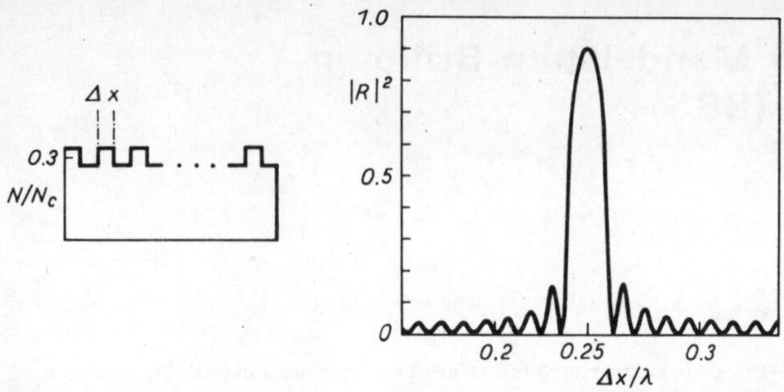

Fig. 9.1 Reflectivity of a spatially modulated density profile (averaged density $N = 0.3\ N_c$) as function of the ratio Δx to the wavelength λ. Number of periods: 50, modulation depth $\pm\ 0.05\ N_c$. The weak oscillations left and right of the maximum are due to the variation of the total plasma length by changing Δx and are present even for a homogeneous profile (cf. Chapter 5).

$l \simeq \lambda/2$, the shape of which depends on both the plasma length and the modulation depth. One should note that the effect does not crucially depend on details of the density variation within the periodicity interval. It is further present also in the case when the density disturbance moves relative to the plasma. In this case, however, the reflected wave suffers from a Doppler shift in frequency.

Stimulated Brillouin scattering results, if the incident wave itself produces a periodic density disturbance with $l \simeq \lambda/2$. This is possible via the ponderomotive force resulting from the superposition of the incident and the reflected wave along the feedback line mentioned above. To initiate the instability, the presence of at least one of the two decay waves with low amplitude is needed at the beginning. Scattering of the incident wave at thermal density fluctuations of proper wavelength is sufficient.

The SBS instability is recognized as important to the progress of laser fusion. By scattering the laser light away from the absorption region near the critical density it may reduce the amount of laser energy available for coupling to the pellet. Therefore, much theoretical and experimental effort was made to understand the behaviour of this instability.

In the various experiments on SBS at laser-plasma interaction /1 - 36/ quite different conditions have been realized with respect to both the incident wave (frequency, pulse length, power, focal spot diameter) and the targets (material, geometry, condition). Quite different are also the observed saturated reflectivities, varying from a few percent up to 60 %. The fractional backscatter often seems to be almost uncorrelated with anything.

In experiments where the generated plasmas have a short scale length of the order of several vacuum wave lengths (short pulses, small focal spot), the SBS reflectivity saturates at a low level as the irradiance increases. On the other hand, long scale length plasmas ($\gtrsim$ 100 vacuum wave lengths) show much more backscatter. Such plasmas can be produced, for example, by using a prepulse to produce the target plasma which then is irradiated by the main laser beam.

Long scale length plasmas can also be produced by a gas jet. Under such conditions Brillouin backscatter was observed to saturate at higher levels (40 % - 60 %).

In experiments with solid targets, an inhomogeneous, expanding plasma with supercritical density regions is produced which gives rise to complications in observing the SBS signal: The scattered light must be separated from the light that is specularly reflected from the critical layer. In case of a plane target this difficulty can be overcome by oblique incidence /10, 14, 21, 30, 34/. The scattered light then mainly retraces the path of the incoming laser beam while the critical layer acts as a mirror. At normal incidence, the separation of backscattered and back-reflected light can only be achieved by a spectral analysis which allows to identify the SBS signal owing to its red shift. Both frequencies, however, may be blue-shifted; the SBS signal due to the Doppler shift produced by the plasma expansion and the specularly reflected signal due to the Doppler shift originating from a possible outward motion of the critical surface, respectively /26, 34/. A blue shift of SBS spectra may also be produced by filamentation /34/.

Complications due to the presence of reflected radiation do not appear when SBS is studied in experiments with undercritical plasmas or gas targets /6, 7, 12, 13, 16, 22, 29/. In this case the plasma is created by a discharge or a second laser and almost uniform conditions can be realized.

In several experiments with high irradiance oscillations in the scattered intensity or the amplitudes of the excited ion acoustic waves have been observed /2, 8, 21, 26, 29/. The period of the oscillations (larger than the ion acoustic period $2\pi / \omega_s$) is almost the same as that observed in harmonic radiation. Stimulated Mandelstam-Brillouin scattering has also been observed in microwave-plasma experiments /37, 38/.

Earlier than experiments, significant effort was made on the theory of stimulated Mandelstam-Brillouin scattering. Gorbunov /39/, utilizing one-dimensional fluid and kinetic descriptions, calculated thresholds and growth rates for backward scattering. The behaviour of the instability in homogeneous /40 - 43/, finite homogeneous /41, 44 - 54/ and inhomogeneous /41, 55 - 59/ plasmas has intensively been studied and is quite well understood. A particular role plays the model of convective SBS from a finite-length homogeneous plasma with heavily damped ion acoustic waves, first introduced by Tang /44, 45/. The influence of, respectively, plasma flow /60 - 68, 72/, reflecting boundaries /69 - 73, 114 - 116/ and a feedback of the scattered wave /74 - 76, 117/ on the SBS instability has also been studied. Particular attention has been paid to various saturation mechanism of the SBS instability, such as linear ion acoustic damping due to ion heating /11, 44, 50, 77, 78/, ion trapping /16, 33, 78 - 80/, the generation of ion acoustic harmonics /81 - 88/ and subharmonics /89 - 91/, wave breaking /13/, momentum deposition /92/, phase mixing /93/ and collisional damping of the high-frequency waves /52/. Limitation of SBS by using multiline lasers or through finite band width effects is studied in /72, 94 - 97/. From the point of view of phase conjugation and ray retracing, stimulated Brillouin scattering has been investigated in /98 - 102/.

A number of numerical experiments on SBS should be mentioned /92, 94, 104 - 110/ which preferentially use a hybrid code, treating the electrons as a fluid and the ions as particles. These simulations are of interest especially for high-intensity fields which cause a strong modification of the particle velocity distribution. In this chapter, we restrict ourselves to the presentation of several aspects of the space-time development of SBS in a uniform plasma of finite length after referring the standard theory of SBS in a homogeneous unbounded plasma.

9.1 SBS in a homogeneous infinite plasma (temporal problem)

In this section we study stimulated Mandelstam-Brillouin scattering within the framework of coherent three-wave interaction in a uniform infinite plasma which provides the simplest theoretical approach to SBS. We represent the electric field as superposition

$$\vec{E}(\vec{r}, t) = \mathrm{Re}\left\{ \vec{E}_1(t)\, e^{i(\vec{k}_1\vec{r} - \omega_1 t)} - i\vec{E}_2(t)\, e^{i(\vec{k}_2\vec{r} - \omega_2 t)} \right\} \tag{9.1}$$

of the pump wave (ω_1, $\vec{k}_1$) and the scattered wave (ω_2, $\vec{k}_2$) and assume the density disturbance to be an ion acoustic wave (ω_s, $\vec{k}_s$):

$$N(\vec{r}, t) = N_o \mathrm{Re}\left\{ q(t)\, e^{i(\vec{k}_s\vec{r} - \omega_s t)} \right\} . \tag{9.2}$$

The interacting waves have to satisfy the matching conditions in frequency and wavenumber

$$\omega_1 = \omega_2 + \omega_s , \tag{9.3}$$

$$\vec{k}_1 = \vec{k}_2 + \vec{k}_s . \tag{9.4}$$

and the dispersion relations

$$k_1^2 c^2 = \omega_1^2 - \omega_p^2 , \tag{9.5}$$

$$k_2^2 c^2 = \omega_2^2 - \omega_p^2 , \tag{9.6}$$

$$k_s^2 c_s^2 = \omega_s^2 . \tag{9.7}$$

The basis system which describes the interaction is given by the pair of Eqs. (8.23), (8.24). In order to simplify the coupling term in the ion acoustic wave equation we introduce a dimensionless electric field, scaled by $\omega m_e v_T/e$. This quantity will be used throughout the following sections. That is, E stands henceforth for $eE/\omega m_e v_T$, the absolute value of which is identical to the ratio v_E/v_T of the electron quiver velocity to the electron thermal velocity. Eqs. (8.23), (8.24) then read

$$2i\omega \frac{\partial \vec{E}}{\partial t} + c^2 \triangle \vec{E} + \omega^2 \mathcal{E}(\omega, N_o)\vec{E} - \omega_p^2 \frac{\delta N}{N_o}\vec{E} = 0 , \tag{9.8}$$

$$\frac{\partial^2 \delta N}{\partial t^2} + 2\sigma \frac{\partial \delta N}{\partial t} - c_s^2 \triangle \delta N = \frac{1}{4} N_o c_s^2 \triangle |\vec{E}|^2 \tag{9.9}$$

where the slowly time-dependent complex amplitude $\vec{E}(\vec{r}, t)$ (subscript ω dropped) is coupled to the real field by Eq. (3.10). After factoring out $\exp(-i\omega_1 t)$, Eq. (9.1) can be written

$$\vec{E}(\vec{r}, t) = \mathrm{Re}\ (\vec{E}_1 e^{i\vec{k}_1\vec{r}} - i\vec{E}_2 e^{i\vec{k}_2\vec{r} + i\omega_s t}) e^{-i\omega_1 t} . \tag{9.10}$$

The expression which now appears as factor at $\exp(-i\omega_1 t)$ is a slowly varying quantity in the sense of Eq. (3.3) and thus can be introduced into the system (9.8), (9.9) together with the ansatz (9.2) for the density disturbance. Using the matching conditions and the dispersion relations (9.3) - (9.7) we get finally the system

$$\frac{d\vec{E}_1}{dt} = -\frac{\omega_1}{4} \frac{N_o}{N_c} q \vec{E}_2 , \tag{9.11}$$

$$\frac{d\vec{E}_2}{dt} = \frac{\omega_1}{4} \frac{N_o}{N_c} q^* \vec{E}_1 , \tag{9.12}$$

$$\frac{dq}{dt} = -\sigma q + \frac{1}{4}\omega_s \vec{E}_1 \vec{E}_2^* \tag{9.13}$$

of ordinary differential equations for the complex amplitudes $\vec{E}_1$, $\vec{E}_2$, q. In deriving this system, weak coupling has been assumed which means that all amplitudes vary slowly during the periods of the waves involved in the interaction. This is a rather stringent condition here, because the amplitudes are especially not allowed to change substantially on the slow ion acoustic time scale $1/\omega_s$. From Eq. (9.12) it follows that the scattered wave has the same polarization as the pump wave. Thus vector notation can be omitted in the mode coupling system. It is further seen that real initial conditions result in all quantities being real for $t > 0$.

Parametric approximation

Setting $E_1(0) = E_o = \text{const.}$ and E_2, $q \propto \exp(\gamma_o t)$, justified during the initial stage of the process, we get from the mode coupling system in the damping-free limit for the growth rate γ_o the expression

$$\frac{\gamma_o}{\omega_1} = \frac{1}{4} \left\{ \frac{N_o}{N_c} \frac{\omega_s}{\omega_1} \right\}^{1/2} E_o . \tag{9.14}$$

To find out, for example, how the growth rate varies with the scattering angle θ, one has to determine the dependence of ω_s on θ. To this end, we rewrite the dispersion relation (9.6) for the scattered wave with the help of the resonance conditions as

$$c^2 (\vec{k}_1 - \vec{k}_s)^2 = (\omega_1 - \omega_s)^2 - \omega_p^2 . \tag{9.15}$$

Using the dispersion relation for the pump wave and neglecting small terms of the order

ω_s/ω_1 we find

$$(\vec{k}_1 - \vec{k}_s)^2 = \vec{k}_1^2 .$$

This is equivalent to

$$\vec{k}_s(2\vec{k}_1 - \vec{k}_s) = 0 \quad \text{or} \quad (\vec{k}_1 - \vec{k}_2)(\vec{k}_1 + \vec{k}_2) = 0 \tag{9.16}$$

which implies $|\vec{k}_1| = |\vec{k}_2|$. Fig. 9.2 (a) illustrates the situation, showing two different vector constellations which satisfy Eq. (9.16). The circle of $r = k_1$ is the locus of the tip of $\vec{k}_2$. Fig. 9.2 (b) shows the resulting polar diagram of the growth rate as function of the scattering angle. The maximum growth rate is attained for backscattering (like SRS). SBS forward scattering does not occur (unlike SRS) because k_s and consequently ω_s go to zero in this case.

In the case of backscattering pump wave and scattered wave have nearly equal but opposite wave vectors. Retaining terms of the order ω_s/ω_1 and using the relation $\omega_1/k_1 = c/\sqrt{\varepsilon}$, $\varepsilon = 1 - N_o/N_c$, from Eq. (9.15) it follows

$$\vec{k}_s = 2\vec{k}_1\left(1 - \frac{1}{\sqrt{\varepsilon}}\frac{c_s}{c}\right) \simeq 2\vec{k}_1 , \tag{9.17}$$

and

$$\vec{k}_2 = \vec{k}_1 - \vec{k}_s = -\vec{k}_1\left(1 - \frac{1}{\sqrt{\varepsilon}}\frac{c_s}{c}\right) \simeq -\vec{k}_1 \tag{9.18}$$

Eq. (9.14) then yields the growth rate

$$\left(\frac{\gamma_o}{\omega_1}\right)_{\text{backsc.}} = \frac{1}{4}\left\{\frac{N_o}{N_c}\, 2\sqrt{\varepsilon}\, \frac{c_s}{c}\right\}^{1/2} E_o \tag{9.19}$$

Fig. 9.2 (c) exhibits the dependence of the backscatter growth rate from the plasma density. The pump-power independent quantity $(\gamma_o/\omega_1)/E_o$ is plotted versus the ratio N_o/N_c .

Expressing the general SBS growth rate (9.14) relative to the ion acoustic frequency we find

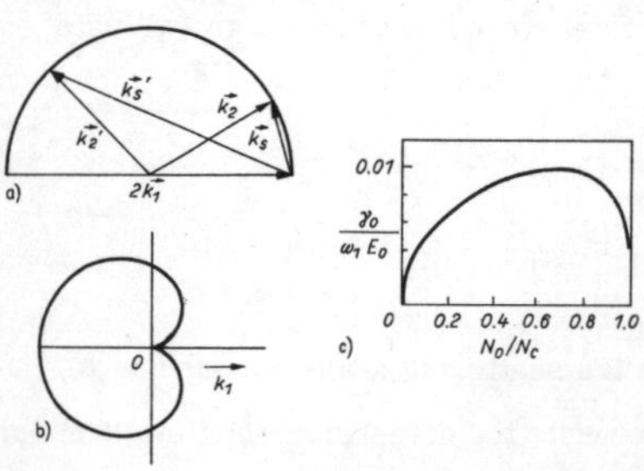

Fig. 9.2 (a) Two vector constellations $\vec{k}_1$, $\vec{k}_2$, $\vec{k}_s$ for stimulated Brillouin scattering in a homogeneous plasma
(b) Polar plot of the SBS growth rate as function of the scattering angle
(c) Normalized SBS growth rate $\gamma_o/(\omega_1 E_o)$ for backscattering versus plasma density

$$\frac{\gamma_o}{\omega_s} = \frac{1}{4} \left\{ \frac{N_o}{N_c} \frac{\omega_1}{\omega_s} \right\}^{1/2} E_o . \qquad (9.20)$$

For typical parameters, e. g., $N_o/N_c = 0.3$, $\omega_s/\omega_1 = 10^{-3}$, and a pump intensity $E_o = 0.1$ this gives $\gamma_o/\omega_s \simeq 0.5$. Thus, the weak-coupling approximation is already violated for moderate pump amplitudes.

Taking into account ion acoustic damping leads to a reduction of the growth rate according to the formula

$$\frac{\gamma_o}{\omega_s} = - \frac{\sigma}{\omega_s} + \left\{ \left(\frac{\sigma}{\omega_s} \right)^2 + \frac{1}{16} \frac{N_o}{N_c} \frac{\omega_1}{\omega_s} E_0^2 \right\}^{1/2} \qquad (9.21)$$

A finite threshold pump intensity results only if both decay waves are damped.

Looking for the development of the SBS instability on a longer time scale requires to take account of the reaction of the growing waves on the pump wave as described by the system (9.11) - (9.13). Taking all quantities to be real and making the further substitutions

$$T = \omega_s t , \quad Q = 4 C q$$

with

$$C = \frac{1}{8} \left(\frac{N_o}{N_c} \frac{2c}{\sqrt{\varepsilon}\, c_s} \right)^{1/2}$$

it can be symmetrized to give

$$\frac{dE_1}{dT} = - C Q E_2 ,$$

$$\frac{dE_2}{dT} = C Q E_1 ,$$

$$\frac{dQ}{dT} = - \frac{\sigma}{\omega_s} Q + C E_1 E_2 . \qquad (9.22)$$

One can easily find the first integral

$$E_1^2 + E_2^2 = \text{const.} \qquad (9.23)$$

and in the damping-free limit the additional invariants

$$E_1^2 + Q^2 = \text{const.} \qquad (9.24)$$

$$E_2^2 - Q^2 = \text{const.} \qquad (9.25)$$

These relations are known as Manley-Rowe relations. From a quantum point of view they may be understood as rules which relate the number of quanta in each mode. If, for example, a quantum disappears from the pump mode, quanta appear in each of the two decay modes. These conservation laws allow to express the solution in terms of elliptic functions. In the case

$$E_1(0) \neq 0, \quad E_2(0) \neq 0, \quad Q(0) = 0$$

we have, for example,

$$E_1(T) = E_1(0)\,\mathrm{sn}(K + CE_3T),$$
$$E_2(T) = E_3\,\mathrm{dn}(K + CE_3T),$$
$$Q(T) = -E_1(0)\,\mathrm{cn}(K + CE_3T) \tag{9.26}$$

where

$$E_3^2 = E_1^2(0) + E_2^2(0)$$

and $K(k)$ is the complete elliptic integral to the modulus $k = E_1(0)/E_3$.

Fig. 9.3 illustrates this case, showing the time evolution of the three wave amplitudes. Similar to the analogous problem at SRS, the periodic character can be understood as alternating operation of Stokes and anti-Stokes scattering. At the time where the pump wave is completely exhausted, the decay waves have maximum amplitudes and thus conditions for the anti-Stokes process are realized. The period is proportional to $K(k)$ and thus goes to infinity for $k \to 1$, i. e. $E_2(0) \to 0$.

The Manley-Rowe relations allow to estimate the maximum density disturbance. From Eq. (9.24) it is seen that $Q^2_{max} = E^2_1(0)$ or

$$q^2_{max} = 2\,\frac{c_s}{c}\,\frac{N_o}{N_c}\sqrt{\mathcal{E}}\;E_o^2. \tag{9.27}$$

For typical parameters ($c_s/c = 10^{-3}$, $N_o/N_c = 0.3$, $E_o < 1$) this expression gives always values much less than one for q^2_{max}. Thus nonlinear effects in the ion acoustic wave cannot affect the time-development of SBS in a homogeneous infinite plasma. We should mention that multiple scattering has been neglected so far. That is, we have not allowed the scattered wave

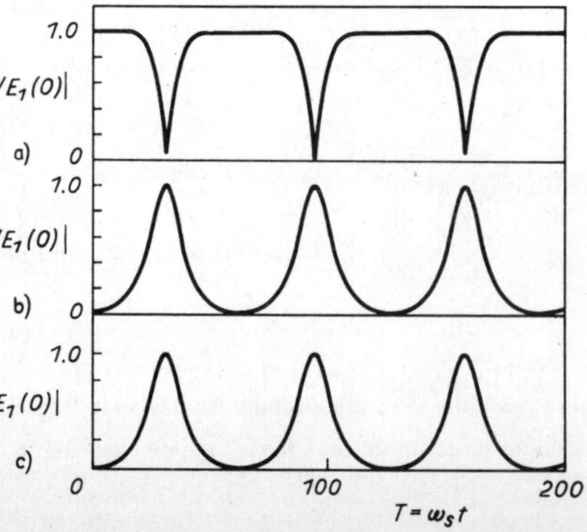

Fig. 9.3 Amplitudes of pump wave (a), scattered wave (b) and ion acoustic wave (c) versus time for SBS backscattering in a homogeneous plasma in the weak coupling limit. $E_1(0) = 0.05$, $N_o/N_c = 0.3$, $c_s/c = 10^{-3}$, $E_2(0) = 10^{-3}$

to act in turn as pump wave producing a scattered wave with the frequency $\omega_1 - 2\omega_s$ and so on. Allowing for multiple scattering we could no longer expect a strictly periodic behaviour of the scattering process in time as found above.

Strong coupling

From Eq. (9.20) we concluded that the weak-coupling approximation breaks down already at pump amplitudes of the order $E_o \gtrsim 0.1$. In order to lift this restriction, the second order time derivative has to be retained in the ion acoustic wave equation. This means that changes of the wave amplitudes on the ion acoustic time scale are considered as slow, being covered by the time dependence of the complex amplitudes $E_1(t)$, $E_2(t)$, $q(t)$. As a consequence, we have to start with the ansatz

$$\vec{E}(\vec{r}, t) = \operatorname{Re}\left\{ (\vec{E}_1(t)\, e^{i\vec{k}_1\vec{r}} - i\vec{E}_2(t)\, e^{i\vec{k}_2\vec{r}})\, e^{-i\omega_1 t} \right\}, \tag{9.28}$$

$$\delta N(\vec{r}, t) = N_o \operatorname{Re}\left\{ q(t)\, e^{i\vec{k}_s\vec{r}} \right\} \tag{9.29}$$

instead of (9.1), (9.2). That is, the ion acoustic disturbance has still the space dependence $\propto e^{i\vec{k}_s\vec{r}}$ but need not to be a wave obeying the dispersion relation (9.7). With other words, there is no definite frequency ω_s that corresponds to k_s and we have no longer to do with a resonant three-wave interaction. Consequently, the resonance condition (9.3) in frequency disappears. While in the weak-coupling limit the position of the vectors $\vec{k}_1$, $\vec{k}_2$, $\vec{k}_s$ is fixed for a given $\vec{k}_1$ and a certain scattering angle, $\vec{k}_s$ now becomes an additional free parameter.

The modified version of the system (9.11) - (9.13) reads

$$\frac{dE_1}{dt} = -\frac{\omega_1}{4}\frac{N_o}{N_c}\, q\, E_2\,, \tag{9.30}$$

$$\frac{dE_2}{dt} = -\frac{ic}{2\omega_1}\left(k_s^2 - 2(\vec{k}_s\vec{k}_1)\right) E_2 + \frac{\omega_1}{4}\, q^* E_1\,, \tag{9.31}$$

$$\frac{d^2q}{dt^2} + 2\sigma\frac{dq}{dt} + k_s^2 c_s^2 = -\frac{i}{2}k_s^2 c_s^2\, E_1 E_2^*\,. \tag{9.32}$$

The additional term in Eq. (9.31) guarantees the scattered wave to satisfy the dispersion relation (9.6) indepently of k_s. In contrast to the weak-coupling limit, the quantities E_1, E_2, q governed by the system (9.30) - (9.32) do not remain real for real initial values.

The early stage of the instability can again be treated by the parametric approximation. Assuming $E_1(t) = E_o = \text{const.}$ and making the ansatz

$$E_2^*,\ q \propto e^{-i\Omega t} \tag{9.33}$$

we get from Eqs. (9.30) - (9.32) the dispersion relation

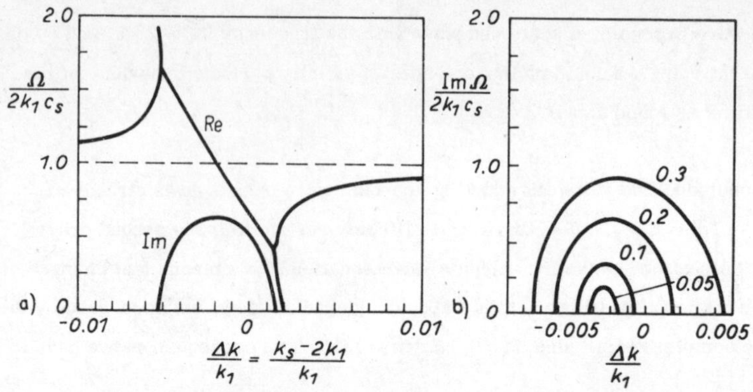

Fig. 9.4 Solutions of the dispersion relation (9.34) resulting from parametric theory of SBS in the case of strong coupling. (a) Re, Im (Ω/k_1c_s) versus $\triangle k/k_1$ for $E_o = 0.2$, $N_o/N_c = 0.3$, $c_s/c = 10^{-3}$; (b) growth of the unstabil k-region with growing pump amplitude (curve parameter), other parameters the same

$$(\Omega^2 + 2i\sigma\Omega - k_s^2 c_s^2)(2\Omega\omega_1 + c^2\vec{k}_s(\vec{k}_s - 2\vec{k}_1)) + \frac{1}{4}k_s^2 c_s^2 \omega_p^2 E_o^2 = 0\,. \quad (9.34)$$

In the pump-free limit this condition reduces to the two dispersion relations for the ion acoustic wave (Ω, k_s) and the scattered wave ($\omega_1 - \Omega$, $k_1 - k_s$).

Fig. 9.4 shows solutions of Eq. (9.34) corresponding to the parameters $c_s/c = 10^{-3}$, $N_o/N_c = 0.3$. In Fig. 9.4 (a) Re Ω and Im Ω are plotted as function of k_s ($\vec{k}_s \parallel \vec{k}_1$) for a pump intensity $E_1(0) = E_o = 0.2$. This figure illustrates how the ion acoustic dispersion is distorted by the pump wave. A certain k_s-band becomes unstable with growth rates reaching values of the order of the real part of the frequency. Fig. 9.4 (b) shows in addition the growth rate for different pump amplitudes. The width of the unstable k_s-region increases with increasing pump intensity. In the limit of small pump amplitudes the weak-coupling approximation is reproduced. This is reflected by the fact that the unstable k-region merges into the point

$$\frac{\triangle k}{k_1} = -\frac{2}{\sqrt{\varepsilon}}\,\frac{c_s}{c}\,, \quad (9.35)$$

which, according to Eq. (9.17), represents the weak-coupling resonance condition for back-scattering.

Fig. 9.5 illustrates the time development of the SBS instability under strong coupling conditions in the damping-free case (backscattering, k_s chosen according to Eq. (9.35) in order achieve maximum growth rate). With growing initial pump amplitude the time evolution changes from periodic to a more and more random behaviour.

Taking account of ion acoustic damping, the oscillatory part of the solution is damped out after a certain time and the configuration eventually reaches a steady state in which a standing spatially periodic density modulation coexists with two counterpropagating electromagnetic

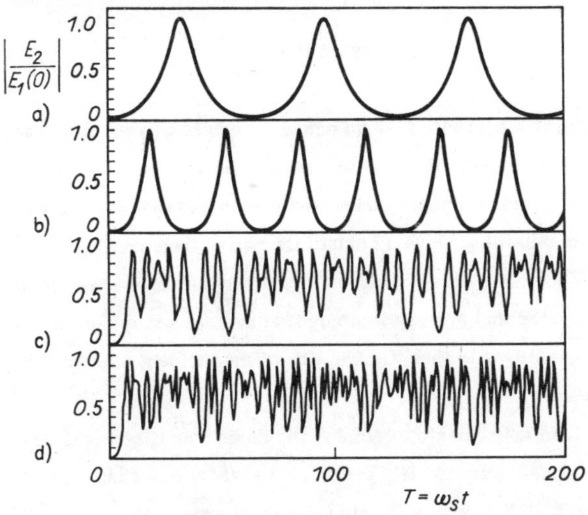

Fig. 9.5 Amplitude of the backscattered wave versus time for SBS in a homogeneous plasma (strong coupling) for increasing initial pump amplitude; E_1 (0) = 0.05 (a), 0.1 (b), 0.2 (c), 0.3 (d); other parameters as in Fig. 9.4

waves of the same frequency but in general different amplitudes. This is illustrated in Fig. 9.6. It shows the time evolution of the three amplitudes (pump wave, scattered wave, ion acoustic disturbance) for different initial pump intensities. The steady-state frequency of the electromagnetic waves is red shifted against the initial pump frequency.

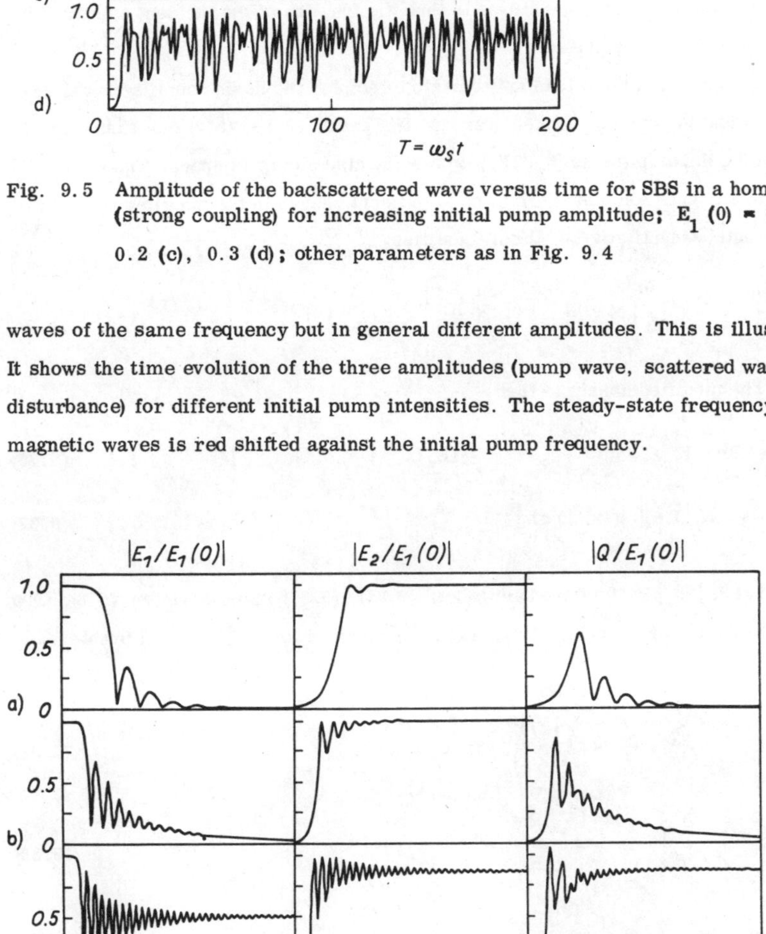

Fig. 9.6 Amplitudes of pump wave, backscattered wave and ion acoustic disturbance versus time for different initial pump intensities in the presence of ion acoustic damping (σ/ω_s = 0.1)

E_1 (0) = 0.1 (a), /.2 (b), 0.3 (c); other parameters as in Fig. 9.4

9.2. Space-time development of SBS in a plasma of finite length

Treating SBS as an initial value problem in an infinite plasma actually results in a simple analysis, the results of which, however, have in general only little relevance to experimental situations. Usually, there exists only a spatially limited scattering region and the pump wave penetrates the plasma from one side, maintaining there a certain intensity, thus creating a spatially unsymmetric configuration. Under this circumstances a spatially homogeneous evolution of the SBS instability cannot be expected and one has to study its development in both space and time. One of the most interesting quantities is then the fraction of the incident energy that is reflected due to the scattering process.

To start the space-time treatment it is appropriate to employ the mode coupling model of the preceding section and to consider backscattering. The ansatz (9.1), (9.2) can easily be generalized by allowing the amplitudes E_1, E_2, q to vary also slowly in space. The term "slowly" means in this connection that there is no significant change in the amplitudes over the distance of a wavelength. Quantitatively, the inequalities

$$\left|\frac{\partial E_{1,2}}{\partial x}\right| \ll k_{1,2} \left|E_{1,2}\right| , \qquad \left|\frac{\partial q}{\partial x}\right| \ll k_s \left|q\right|$$

must be satisfied. The modified ansatz is then

$$E(x, t) = \mathrm{Re}\left[E_1(x, t)\, e^{ik_1 x - i\omega_1 t} - iE_2(x, t)\, e^{ik_2 x - i\omega_2 t}\right], \tag{9.36}$$

$$\delta N(x, t) = N_o \,\mathrm{Re}\left[q(x, t)\, e^{ik_s x - i\omega_s t}\right], \tag{9.37}$$

where we suppose that $k_1, k_s > 0$. Introducing Eqs. (9.36), (9.37) into the system (9.8), (9.9) and going through the same sort of calculations as just performed for the temporal problem we arrive at the system

$$\frac{\partial E_1}{\partial t} + \sqrt{\varepsilon}\, c \frac{\partial E_1}{\partial x} = -\frac{\omega_1}{4} \frac{N_o}{N_c}\, q\, E_2 , \tag{9.38}$$

$$\frac{\partial E_2}{\partial t} - \sqrt{\varepsilon}\, c \frac{\partial E_2}{\partial x} = \frac{\omega_1}{4} \frac{N_o}{N_c}\, q^* E_1 , \tag{9.39}$$

$$\frac{\partial q}{\partial t} + c_s \frac{\partial q}{\partial x} = -\sigma q + \frac{1}{4}\omega_s E_1 E_2^* \tag{9.40}$$

of partial differential equations for the amplitudes E_1, E_2, q. The coefficient at the space derivative in each equation is the group velocity of the corresponding wave. We suppose that the pump wave is impinging from the left onto a plasma of finite length L. Appropriate boundary conditions are then

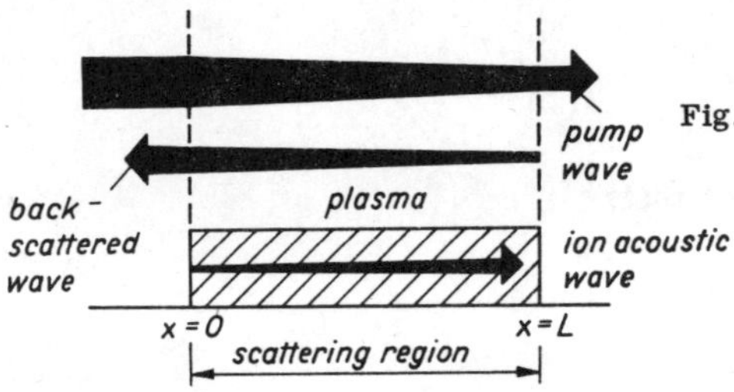

Fig. 9.7 Scheme of stimulated Brillouin backscattering in a homogeneous plasma of finite extent

$$q(0, t) = 0, \qquad E_2(L, t) = 0 \tag{9.41}$$

They express the plausible requirements that both the ion acoustic wave and the scattered wave start propagating with zero amplitudes from the edges of the plasma. The situation is sketched in Fig. 9.7.

Parametric theory

For a first approach it is appropriate to look for solutions of the mode coupling system (9.38) - (9.40) under the condition of a constant pump wave $E_1(x, t) = E_o = $ const. We are then left with the two linear equations

$$\frac{\partial E_2}{\partial t} - \sqrt{\varepsilon}\, c \frac{\partial E_2}{\partial x} = \frac{\omega_1}{4} \frac{N_o}{N_c} E_o q^*, \tag{9.42}$$

$$\frac{\partial q}{\partial t} + c_s \frac{\partial q}{\partial x} = -\sigma q + \frac{1}{4} \omega_s E_o E_2^* \tag{9.43}$$

the solutions of which can be analyzed in terms of the theory developed in the Appendix. Introducing the ansatz E_2^*, $q \propto \exp i(Kx - \Omega t)$ results in a dispersion relation $D(\Omega, K) = 0$. Solutions that fit the boundary conditions (9.41) require a superposition

$$E_2(x, t) = (A e^{iK_1 x} + B e^{iK_2 x}) e^{-i\Omega t} \tag{9.44}$$

where K_1, K_2 are solutions of $D(\Omega, K) = 0$. Applying the results of the Appendix we find that the boundary conditions are met only when the eigenvalue equation

$$\cos \frac{\pi}{2} P - i \frac{(F + iS)}{P} \sin \frac{\pi}{2} P = 0 \tag{9.45}$$

is satisfied. Here, the following notations have been introduced:

$$F = \frac{1}{\pi} \frac{L\Omega}{c_s} \left(1 + \frac{c_s}{c}\right) \simeq \frac{1}{\pi} \frac{L\Omega}{c_s} = 4\sqrt{\varepsilon} \frac{L}{\lambda_o} \frac{\Omega}{\omega_s}, \tag{9.46}$$

$$S = \frac{1}{\pi} \frac{L\sigma}{c_s} = 4\sqrt{\varepsilon} \frac{L}{\lambda_o} \frac{\sigma}{\omega_s} = \frac{L}{L_\sigma}, \tag{9.47}$$

$$Y = (2N_o/N_c)^{1/2} \frac{L}{\lambda_o} E_o = \frac{L}{L_o}, \tag{9.48}$$

$$P = \left\{ (F + iS)^2 + Y^2 \right\}^{1/2} . \tag{9.49}$$

It is convenient to interpret, respectively, the quantities

$$\left(\frac{N_c}{2N_o} \right)^{1/2} \frac{\lambda_o}{E_o} = L_o \tag{9.50}$$

as a basic gain length and

$$\frac{\pi c_s}{\sigma} = \frac{\lambda_o}{4\sqrt{\varepsilon}} \frac{\omega_s}{\sigma} = L_\sigma \tag{9.51}$$

as a generalized damping length. Neglecting c_s/c against one in expression (9.46) is equivalent to ignore the time derivative $\partial E_2/\partial t$ in Eq. (9.42) for the scattered wave.

Eq. (9.45) admits only purely imaginary solutions for the generalized frequency F. Since F enters Eq. (9.45) only in the combination $Z = F + iS$, it is convenient to display these solutions as $\mathrm{Im}\, Z = \Gamma + S$ ($F = i\Gamma$, Γ normalized growth rate) versus the generalized pump amplitude Y (Fig. 9.8). There exists a multiple spectrum of eigensolutions, emerging from the points $Y = 2m + 1$, $m = 0, 1, 2, \ldots$ on the Y-axis. The corresponding spatial profiles differ in geometry. While the fundamental mode ($m = 0$) is characterized by monotoneous profiles, the non-fundamental modes ($m \geq 1$) exhibit zeros in the amplitude profiles, the number of which grows with m. The graphs in Fig. 9.8 describe how the growth rate varies with Y. Each branch has its own threshold, in absence of damping given by

$$Y_{thr}^{(m)} = 2m + 1 . \tag{9.52}$$

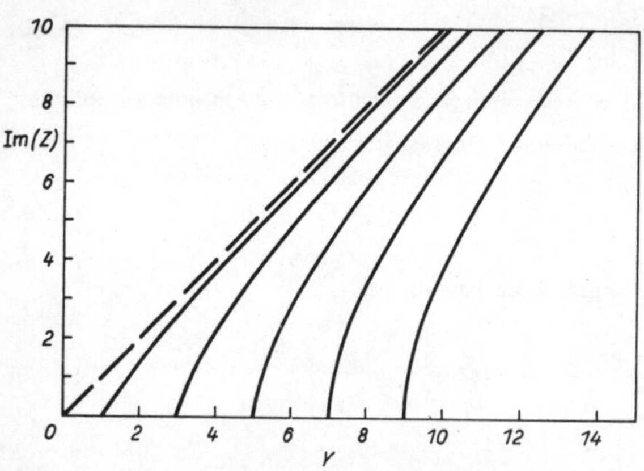

Fig. 9.8 The purely imaginary solutions of Eq. (9.45) in the form $\mathrm{Im}\, Z = \Gamma + S$ (Γ normalized growth rate, S generalized ion acoustic damping decrement) versus generalized pump amplitude Y. The different branches correspond to different spatial amplitude profiles

Thus, the lowest threshold is $Y_{thr} = 1$. Hence, in a plasma without ion acoustic damping the condition for an absolute SBS instability to occur is

$$Y = L/L_o = (2N_o/N_c)^{1/2} \frac{L}{\lambda_o} E_o > 1 , \tag{9.53}$$

i. e., the scattering length must exceed the gain length L_o. Since scattering length and pump amplitude enter the formula via the product $L \cdot E_o$, a low pump field strength can be compensated by a large scattering length to exceed the threshold in a plasma free of ion acoustic losses and vice versa.

In the presence of ion acoustic damping the threshold is shifted to higher values. This can simply be concluded from the graphs in Fig. 9. 8 when interpreted in a modified way. The threshold amplitude Y_{thr} as function of the generalized damping decrement S is obviously given by inversion of the dependencies shown in Fig. 9. 8. This is illustrated in Fig. 9. 9 for the branch with the lowest threshold. As may be seen, in the limit $S \rightarrow 0$ the threshold approaches the damping-free limit $Y_{thr} = 1$. Y_{thr} can simply be expressed analytically as function of S in the cases $S \ll 1$ and $S \gg 1$. In the former case we have

$$Y_{thr} = 1 + S , \quad S \ll 1 .$$

Rewritten as a condition for the pump amplitude we find

$$E_{o,thr} = \frac{L}{\lambda_o} \left(2 \frac{N_c}{N_o} \right)^{1/2} + 2 \left(2\varepsilon \frac{N_c}{N_o} \right)^{1/2} \frac{\sigma}{\omega_s} \tag{9.54}$$

In the opposite case we get

$$Y_{thr} \simeq S , \quad S \gg 1$$

i. e.,

$$E_{o,\,thr} = 2 \left(2\,\varepsilon \frac{N_c}{N_o} \right)^{1/2} \frac{\sigma}{\omega_s} \tag{9.55}$$

$S \gg 1$ can be achieved by strong ion acoustic damping and/or a large scattering region. Eq. (9.55) indicates that in this case the threshold pump amplitude no longer depends on the scatter-

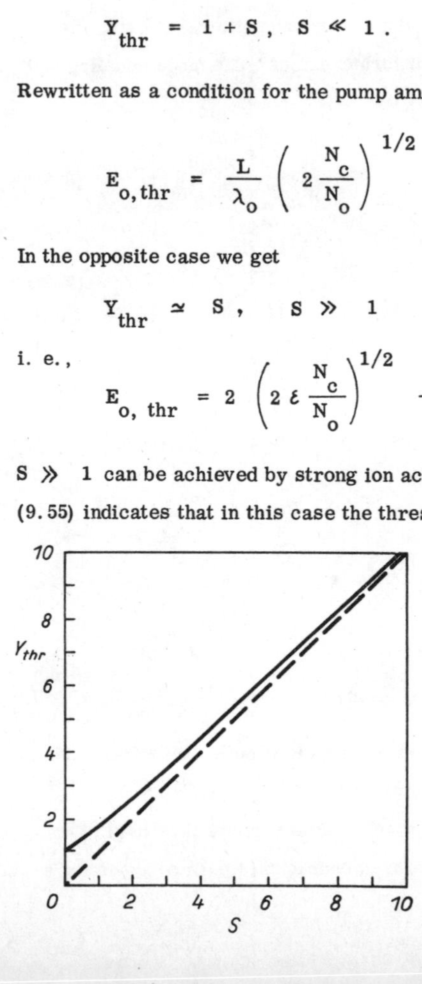

Fig. 9.9 Lowest SBS threshold Y_{thr} versus generalized ion acoustic damping decrement S

ing length L. But even then an absolute SBS instability is possible for sufficiently high pump power.

From Fig. 9.9 it is seen that the inequality $Y_{thr} > S$ is always satisfied. Hence, a necessary condition for the generation of an absolute SBS instability in a plasma with ion acoustic damping is

$$Y > S , \quad \text{i. e.,} \quad L_o > L . \tag{9.56}$$

That is, the gain length must exceed the generalized damping length. This criterion tends to become sufficient in character with growing damping.

Mode coupling analysis

The parametric theory indicated that the term $\partial E_2/\partial t$ in the equation for the scattered wave does not very much influence the development of the SBS instability because of the extremely different group velocities of ion acoustic wave and scattered wave. Obviously, the same is true for the time derivative $\partial E_1/\partial t$ in the equation for the pump wave. Physically, the neglect of these two terms means that the time scale for the development of the instability is dominated by that of the ion acoustic wave and that the electromagnetic amplitude pattern follows adiabatically the changes in the density structure With other words, the electromagnetic field configuration at a certain time is regarded as independent from that at former times. The mode coupling system in reduced form is then

$$\frac{\partial E_1}{\partial x} = - \frac{\omega_1}{4\sqrt{\varepsilon}\, c} \frac{N_o}{N_c} q \, E_2 , \tag{9.57}$$

$$\frac{\partial E_2}{\partial x} = - \frac{\omega_1}{4\sqrt{\varepsilon}\, c} \frac{N_o}{N_c} q^{*} E_1 , \tag{9.58}$$

$$\frac{\partial q}{\partial t} + c_s \frac{\partial q}{\partial x} = - \sigma q + \frac{1}{4} \omega_s E_1 E_2^{*} . \tag{9.59}$$

From Eqs. (9.57), (9.58) it follows immediately that

$$\frac{\partial}{\partial x}\left(|E_1|^2 - |E_2|^2 \right) = 0 ,$$

i. e., at each time the conservation law

$$|E_1|^2 - |E_2|^2 = \text{const.} \tag{9.60}$$

is satisfied.

Fig. 9.10 illustrates the space-time development of stimulated Brillouin scattering in the case of no ion acoustic damping for the choice of parameters $L = 20\ \lambda_o$, $N_o/N_c = 0.3$, $E_o = v_E^{(pump)}/v_T = 0.2$. These parameters combine to give a value Y above threshold. The system (9.57) - (9.59) has been solved applying the boundary conditions (9.41) and assuming a

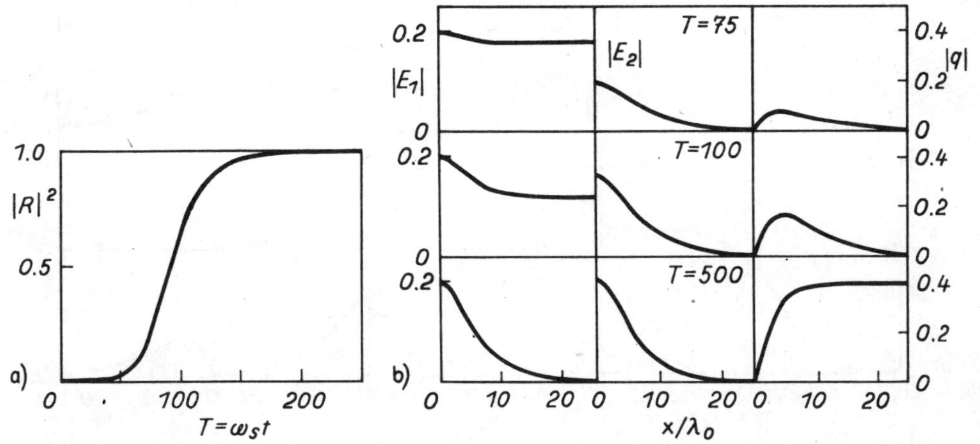

Fig. 9.10 (a) SBS reflectivity versus time; (b) spatial amplitude profiles of pumpwave, backscattered wave and ion acoustic wave at three times. $E_o = 0.2$, $L/\lambda_o = 25$, $N_o/N_c = 0.3$, $c_s/c = 10^{-3}$, $\sigma = 0$; ion acoustic initial level $q_o = 10^{-3}$

noise level $|q| = 0.001$ of the ion acoustic wave at the beginning. The operation regime of the instability is seen to be confined to the left-hand side of the scattering region where the pump wave penetrates the plasma. After a growth of the decay wave amplitudes according to the linear parametric theory, pump wave depletion sets in and the instability develops toward a saturated steady state, even without ion acoustic damping. After leaving the instability region, the ion acoustic wave propagates with constant amplitude. The steady-state solution neither depends on the initial ion acoustic noise level nor needs non-zero boundary values q (o, t), E_2 (L, t) to persist. This reflects the absolute character of the instability. Obviously, the basic features of the SBS evolution in time differ substantially from those in both space and time.

If ion acoustic damping is present, we find a somewhat modified evolution. An example is illustrated in Fig. 9.11 for a damping decrement of $\sigma / \omega_s = 0.025$, other parameters the same as in Fig. 9.10. The most notable modifications are a reduced reflectivity and a decrease of the ion acoustic amplitude outside the instability region. Obviously, an extention of the scattering length in this case would not result in an increase in the steady-state reflection.

Numerical solutions of the mode coupling equations confirm the predictions of the parametric theory concerning the existence of threshold pump amplitudes and their dependence on the ion acoustic damping decrement.

Finally, we point to a formal analytical solution of the mode coupling system (9.57) - (9.59) which gives additional insight in the scattering process. For a given ion acoustic amplitude q(x, t) within the scattering interval $0 \leq x \leq L$ a solution is provided by the ansatz

$$E_1 (x, t) = a e^{\psi} + b e^{-\psi} , \tag{9.61}$$

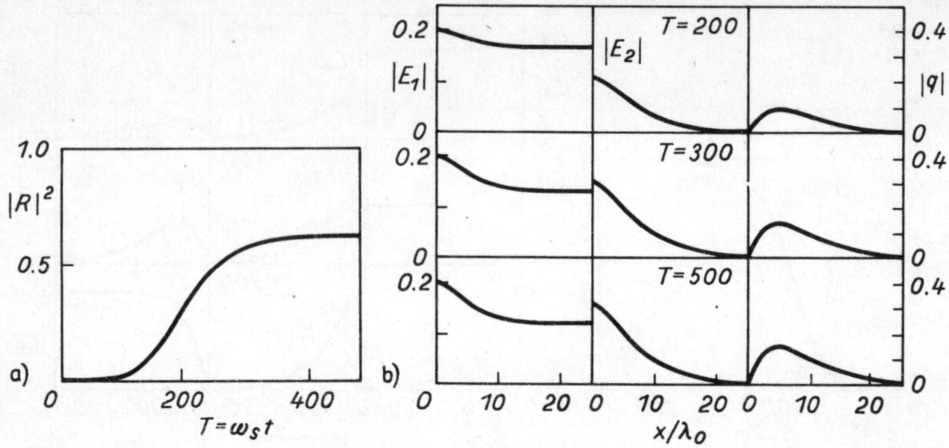

Fig. 9.11 The same as Fig. 9.10, but presence of ion acoustic damping ($\sigma / \omega_s = 0.025$)

$$E_2(x, t) = -a e^{\psi} + b e^{-\psi} \tag{9.62}$$

where

$$\psi(x, t) = -\frac{\omega_1}{4\sqrt{\varepsilon'}\, c} \frac{N_o}{N_c} \int_o^x q(x', t)\, dx'$$

The coefficients a, b are determined by the boundary conditions. Assuming a constant incident pump amplitude

$$E_1(0, t) = E_o = \text{const.}$$

and zero scattering intensity at the right-hand boundary, i. e.,

$$E_2(L, t) = 0,$$

we find

$$E_1(x, t) = E_o \frac{\cosh(\psi - \psi_L)}{\cosh \psi_L}, \tag{9.64}$$

$$E_2(x, t) = -E_o \frac{\sinh(\psi - \psi_L)}{\cosh \psi_L} \tag{9.65}$$

where ψ_L is given by the relation

$$\psi_L = \psi(L, t) = -\frac{\omega_1}{4\sqrt{\varepsilon}\, c} \frac{N_o}{N_c} \int_o^L q(x, t)\, dx. \tag{9.66}$$

Owing to the identity $\cosh^2(z) - \sinh^2(z) = 1$ the conservation law (9.60) is satisfied . The reflection coefficient associated with the given density disturbance q (x, t) can easily be expressed as

$$R(t) = \frac{E_2(0, t)}{E_o} = \tanh \psi_L , \tag{9.67}$$

the corresponding energy reflection is given by $|R(t)|^2$. Formula (9.67) reveals the interesting fact that R may vanish without the ion acoustic amplitude being indentically zero in the interval (0, L). Zero reflectivity is attained if

$$\int_0^L q(x, t)\, dx = 0 . \tag{9.68}$$

This condition can be met, for example, if q undergoes a proper phase change along the scattering path.

The ion acoustic amplitude shown in Fig. 9.10 grows up to a stationary level of the order $|q| \simeq 0.5$ at an incident pump power of $E_o = 0.2$. This indicates that the linearized treatment of the ion sound wave breaks down already at moderate pump amplitudes, in contrast to the purely temporal problem. On the other hand, the weak-coupling approximation is still applicable for pump amplitudes $E_o \gtrsim 0.2$ owing to the significantly reduced growth rates.

9.3 SBS as steady-state spatial problem

In the preceding section it was shown that in the presence of a sufficiently strong pump wave the SBS instability developes in space and time from an initial ion acoustic noise toward a steady-state solution. This saturated state is characterized by pump wave depletion and a certain non-zero reflectivity and appeared to be nonperiodic in space. In this section we examine this stationary solution, ignoring the transient behaviour and treating SBS as a purely spatial problem. The most interesting question is, how the steady-state reflectivity depends on the parameters entering the problem (e. g., scattering length L, pump amplitude E_o, ion acoustic damping decrement σ). The corresponding mode coupling system results from Eqs. (9.57) - (9.59) by cancelling the time derivative in the equation for the ion acoustic wave:

$$\frac{dE_1}{dx} = -\frac{\omega_1}{4\sqrt{\varepsilon}\, c} \frac{N_o}{N_c}\, q\, E_2 , \tag{9.69}$$

$$\frac{dE_2}{dx} = -\frac{\omega_1}{4\sqrt{\varepsilon}\, c} \frac{N_o}{N_c}\, q^* E_1 , \tag{9.70}$$

$$\frac{dq}{dx} = -\frac{\sigma}{c_s}\, q + \frac{1}{4} k_s E_1 E_2^* . \tag{9.71}$$

Substituting

$$z = \frac{\pi}{2}\frac{x}{L}, \tag{9.72}$$

$$Q = \left(\frac{1}{2\mathcal{E}}\frac{N_o}{N_c}\right)^{1/2} q \tag{9.73}$$

the system can be cast into a symmetric form

$$\frac{dE_1}{dz} = -C\,Q\,E_2,$$

$$\frac{dE_2}{dz} = -C\,Q\,E_1,$$

$$\frac{dQ}{dz} = -2\,S\,Q + C\,E_1\,E_2 \tag{9.74}$$

which appears to be the spatial counterpart of the system (9.22). Here, E_1, E_2, Q are supposed to be real which is justified in the case of real boundary values. $S = L/L_\sigma$ is the generalized ion acoustic damping decrement, defined by Eq. (9.47), and C is given by

$$C = (2N_o/N_c)^{1/2}\,\frac{L}{\lambda_o}. \tag{9.75}$$

The system (9.74) possesses the invariant

$$E_1^2 - E_2^2 = \text{const} \tag{9.76}$$

and in absence of ion acoustic damping the two additional integrals

$$E_1^2 + Q^2 = \text{const},$$

$$E_2^2 + Q^2 = \text{const}. \tag{9.77}$$

These spatial Manley-Rowe relations differ in signs from their temporal analogs. Again, the solution can be expressed by Jacobi elliptic functions. Assuming the boundary values

$$E_1(0) = E_o \neq 0,\quad E_2(0) \neq 0,\quad Q(0) = 0 \tag{9.78}$$

at the left-hand boundary we find, written as function of x, the solutions

$$E_1(x) = E_o\,\mathrm{dn}\left(\frac{\pi}{2}\frac{x}{L_o}\right),$$

$$E_2(x) = E_2(0)\,\mathrm{cn}\left(\frac{\pi}{2}\frac{x}{L_o}\right),$$

$$Q(x) = E_2(0)\,\mathrm{sn}\left(\frac{\pi}{2}\frac{x}{L_o}\right). \tag{9.79}$$

L_o is the gain length, introduced previously by Eq. (9.50). The modulus k is given by

$$k = E_2(0)/E_o = R, \tag{9.80}$$

being identical to the reflection coefficient. The solution (9.79) is not yet determined completely. We have in addition to impose a boundary condition on the scattered wave at the right-hand boundary. Requiring $E_2 = 0$ at $x = L$ leads to the condition

$$\mathrm{cn}\left(\frac{\pi}{2}\frac{L}{L_o}\right) = \mathrm{cn}\left(\frac{\pi}{2}Y\right) = 0. \tag{9.81}$$

The zeros of $\mathrm{cn}(\lambda)$ are located at $\lambda_m = (2m+1)K(k)$, $m = 1, 2, \ldots$, where $K(k)$ is the complete elliptic integral. Thus, we arrive at the additional condition

$$\frac{\pi}{2}Y = (2m+1)K(R). \tag{9.82}$$

This equation is an implicit relation for the SBS reflectivity R associated with the solution (9.79) as function of the pump amplitude E_o, the scattering length L and the relative density N_o/N_c in a plasma without ion acoustic damping. Fig. 9.12 displays solutions of Eq. (9.82) in the form R^2 versus Y for $m = 0, 1, 2$. Since $K(0) = \pi/2$, the marginal existence condition for a solution corresponding to the branch m is given by

$$Y = 2m+1 = Y^{(m)}.$$

Thus, a non-trivial steady-state solution belonging to the branch m exists only if

$$Y > Y^{(m)} = 2m+1.$$

This condition is identical to the threshold condition (9.52) predicted by the parametric theory.

The points $Y = 3, 5$ in Fig. 9.12 (a) represent parameter combinations where zero reflectivity is attained without the density disturbance being identically zero within the scattering region. Fig. 9.12 (b) - (d) shows spatial amplitude profiles associated with solutions of Eq. (9.82) for $m = 0, 1, 2$ in the case of finite reflection. The profiles exhibit a growing number of zeros with growing m. The solutions corresponding to $m \neq 0$, however, play only a singular role, because they are not accessible in a time evolution /47, 49/.

In the limit $L \longrightarrow \infty$ (transition to a plasma half space) the solution (9.79) degenerates to a hyperbolic secant/tangent type

$$\begin{aligned} E_1(x) &= E_2(x) = E_o \,\mathrm{sech}\left(\frac{\pi}{2}\frac{x}{L_o}\right), \\ Q(x) &= E_o \tanh\left(\frac{\pi}{2}\frac{x}{L_o}\right) \end{aligned} \tag{9.83}$$

which goes along with a transition to complete reflection of the incident wave.

In the presence of ion acoustic damping the system (9.74) cannot longer be solved exactly in terms of Jacobi elliptic functions. One is confronted with a two-point boundary problem which can numerically be solved by a shooting procedure.

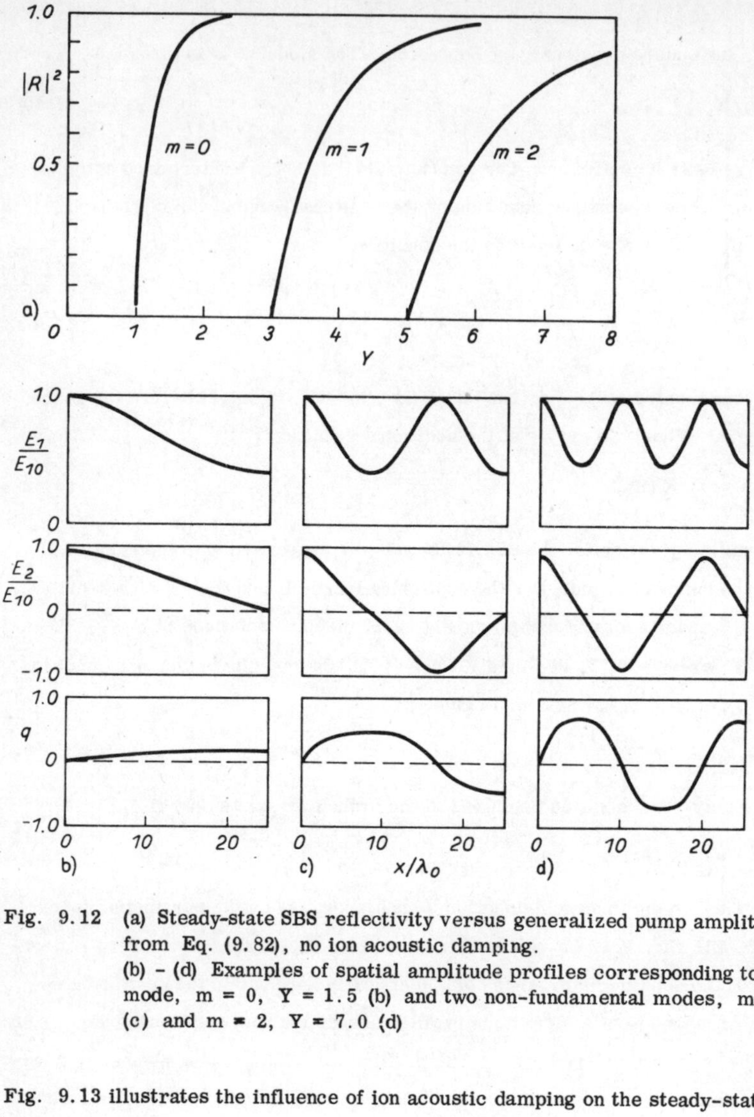

Fig. 9.12 (a) Steady-state SBS reflectivity versus generalized pump amplitude Y, calculated from Eq. (9.82), no ion acoustic damping.
(b) – (d) Examples of spatial amplitude profiles corresponding to the fundamental mode, m = 0, Y = 1.5 (b) and two non-fundamental modes, m = 1, Y = 4.5 (c) and m = 2, Y = 7.0 (d)

Fig. 9.13 illustrates the influence of ion acoustic damping on the steady-state SBS reflectivity. Keeping the scattering length constant the graphs describe the variation of the reflectivity with pump amplitude (curve parameter: S). Saturation occurs in this case always at total reflection.

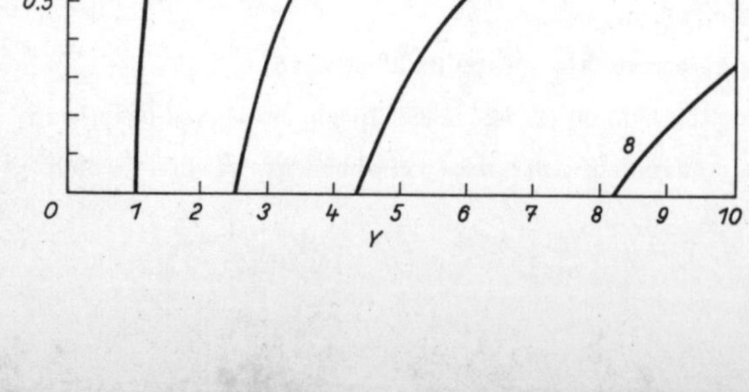

Fig. 9.13 Steady-state SBS reflectivity as function of the generalized pump amplitude Y for several values of the generalized ion acoustic damping decrement S (curve parameter)

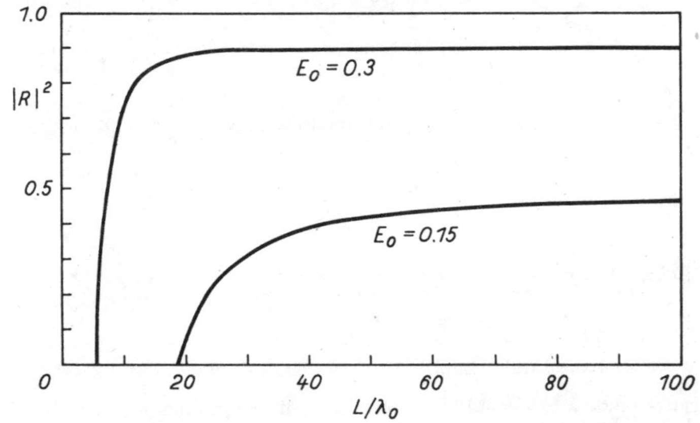

Fig. 9.14 Steady state SBS reflectivity in the presence of ion acoustic damping as function of the scattering length for two pump intensities ; $N_o/N_c = 0.3$, $c_s/c = 10^{-3}$, $\sigma/\omega_s = 0.025$.

For pump amplitudes $E_o \lesssim 0.1$ the threshold for the absolute instability cannot longer be exceeded in this case even for large L because criterion (9.56) is violated.

The situation changes, however, when varying the scattering length at constant pump amplitude because the generalized damping decrement S also contains L. Fig. 9.14 shows an example which indicates saturation at a value below one for $L \to \infty$ in this case.

9.4 Convective SBS instability

A steady-state SBS reflectivity can be maintained even for zero boundary values q (0), E_2 (L) when the threshold for an absolute instability (Fig. 9.9) is exceeded. For parameter combinations below threshold, there is no instabil behaviour and the reflection coefficient keeps being zero. This strict threshold behaviour is weakened when a steady source exists in the system, represented, for example, by non-zero boundary values of the ion acoustic disturbance at the side where the pump wave is impinging or the scattered wave at the opposite side, respectively. Finite boundary values q (o), E_2 (L) give rise to an instability even for sub-threshold conditions but the subsequently attained steady-state reflectivity depends crucially on the boundary values and goes to zero together with them. This is the characteristic feature of a convective instability.

The two boundary values influence the solution in a different way. A finite value of the ion acoustic amplitude q (0) does not seriously affect the instability domain. This is demonstrated in Fig. 9.15 which compares the dependencies $|R|^2$ as function of Y for the two cases $q(0) = q_o = 0$ and $q_o = 0.01$. Stationary SBS reflectivities are seen to occur also for values $Y < 1$, i. e., below threshold.

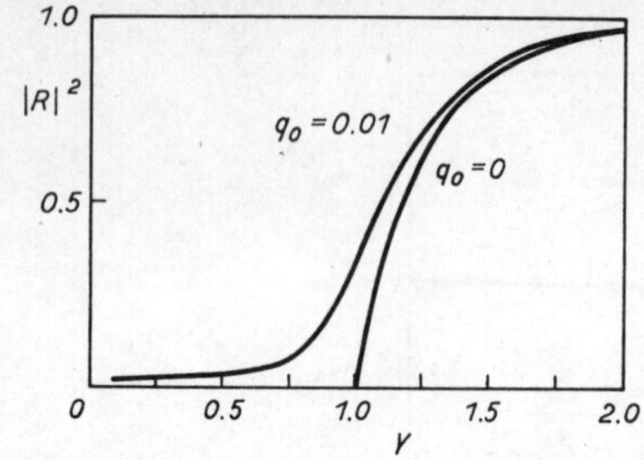

Fig. 9.15 Steady state SBS reflectivity versus generalized pump amplitude Y assuming a non-zero ion acoustic amplitude level $q_o = 0.01$ at the left-hand boundary. For comparison also the graph for $q_o = 0$ is shown

A finite boundary value of the scattered wave amplitude, however, can cause a substantial extention of the parameter range leading to instability. Especially, in the presence of ion acoustic damping parameter combinations below threshold are capable of producing a significant reflection level due to a finite $E_2(L)$. While, for example, the reflection coefficient saturates at $|R|^2 < 1$ with growing scattering length in the case of an absolute instability, as shown in Fig. 9.14, the existence of a finite boundary value $E_2(L)$ combined with non-zero ion acoustic damping leads to a monotoneous increase of $|R|^2$ with L. This is illustrated in Fig. 9.16 which shows $|R|^2$ as function of L for the choice of parameters $E_o = 0.1,\ 0.2$, $\sigma/\omega_s = 0.05$, $E_2(L) = 10^{-3}$. Note that condition (9.56) is violated independently of the scattering length for this parameters, i. e., no absolute instability can occur.

Tang's solution

Convective amplification of electromagnetic background noise ($E_2(L) \neq o$) can approximately be treated analytically by neglecting the spatial derivative in Eq. (9.71) for the ion acoustic wave. This gives

$$q = \frac{1}{4}\ \frac{\omega_s}{\sigma}\ E_1 E_2^{*} \ . \tag{9.84}$$

Introducing this expression into the remaining equations for E_1, E_2 we get the system

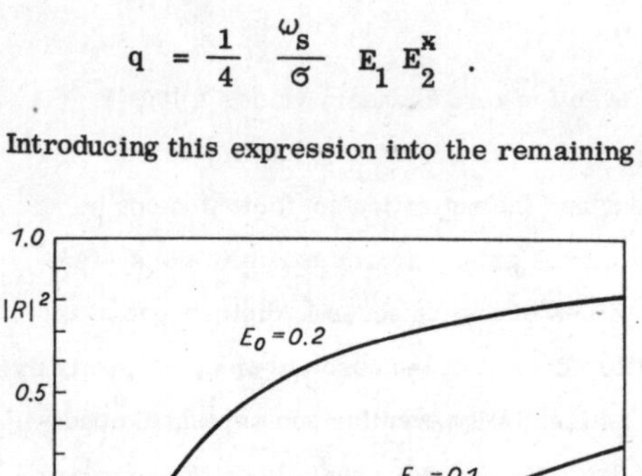

Fig. 9.16 Steady-state SBS reflectivity versus scattering length for two pump amplitudes assuming a finite boundary value ($E_2(L) = 10^{-3}$) of the scattered wave; other parameters: $N_o/N_c = 0.3$, $c_s/c = 10^{-3}$, $\sigma/\omega_s = 0.05$. According to the criterion (9.56) sub-threshold conditions are realized

$$\frac{dE_1}{dx} = -K \left| E_2 \right|^2 E_1 , \tag{9.85}$$

$$\frac{dE_2}{dx} = -K \left| E_1 \right|^2 E_2 \tag{9.86}$$

where

$$K = \frac{\omega_1}{16\sqrt{\varepsilon}\, c} \frac{N_o}{N_c} \frac{\omega_s}{\sigma} .$$

Multiplying Eq. (9.85) by E_1^* and Eq. (9.86) by E_2^* and combining both equations yields

$$\frac{d\left| E_1 \right|^2}{dx} = \frac{d\left| E_2 \right|^2}{dx} = -K \left| E_1 \right|^2 \left| E_2 \right|^2 . \tag{9.87}$$

Introducing the notation

$$I_{1,2} = \left| E_{1,2} \right|^2$$

we immediately find from Eq. (9.87) the conservation law

$$I_1 - I_2 = \text{const} = C > 0 ,$$

and further,

$$\frac{dI_1}{dx} = -K I_1 (I_1 - C) . \tag{9.88}$$

This equation admits the solution

$$I_1(x) = \frac{A}{e^{-KCx} + D}$$

with $C = A/D$. This relation, together with two boundary conditions fix the open parameters A, C, D. It is convenient to write the solution as

$$I_1(x) = I_1(0) \frac{R^2 - 1}{R^2 \exp\left\{ -KI_1(0)\,(1 - R^2)\, x \right\} - 1} , \tag{9.89}$$

$$I_2(x) = I_1(0) \frac{R^2 (R^2 - 1)}{\exp\left\{ KI_1(0)\,(1 - R^2)\, x \right\} - R^2} \tag{9.90}$$

where $R^2 = I_2(0)/I_1(0)$. In this form the solution has been given by Tang /45/. Fig. 9.17 compares Tang's solution with that resulting from the mode coupling system (9.69) - (9.71) for a particular set of parameters ($E_o = 0.2$, $L = 100\ \lambda_o$, other parameters as in Fig. 9.16). The agreement is not quite satisfactorily, especially Tang's solution fails in describing the behaviour near $x = 0$.

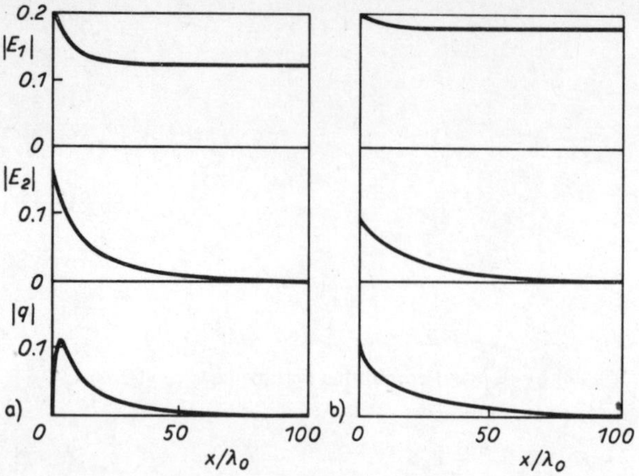

Fig. 9.17 Spatial amplitude profiles of pump wave, backscattered wave and ion acoustic wave for a convective SBS instability maintained by a finite boundary value $E_2(L) = 10^{-3}$ according to the mode coupling system (9.69) - (9.71) (a) and in Tang's approximation (b); $E_o = 0.2$, $L = 100\ \lambda_o$, other parameters as in Fig. 9.16

9.5 Brillouin backscatter pulsations

The results of the previous section indicated that the behaviour of the convective SBS instability sensitively depends on what conditions are realized for the scattered wave at the right-hand boundary of the plasma slab. Stimulated Brillouin backscatter was seen to be initiated, for example, by a small but finite value $E_2(L)$ even for parameter combinations which do not allow the evolution of an absolute instability. In this section we shall examine what consequences could be expected with respect to the time development of the backscatter when the boundary value $E_2(L, t)$ is no longer determined by the electromagnetic noise level, but by a feedback of the scattered wave /74 - 76/. A feedback, for example, is established when the boundary values of the scattered wave at both edges of the scattering region are linked via the relation

$$E_2(L, t) = f\, E_2(0, t) \qquad (9.91)$$

with f as feedback strength ($|f| < 1$). This kind of feedback will be produced by partial double reflection of the scattered wave at the plasma boundaries owing to differences in the refractive indices of the adjacent media. Partial reflection of the backscattered wave at the left-hand boundary first creates a wave propagating parallel to the pump wave with constant amplitude because of no resonant interaction with the other waves. Another partial reflection of the wave at the right-hand boundary results in a finite amplitude $E_2(L, t)$ which is related to $E_2(0, t)$ by an equation of type (9.91). A scheme of this process is shown in Fig. 9.18. For the calculation of the feedback factor we refer to Fig. 9.18 (b), which separates all waves having the frequency of the backscattered wave. The back-travelling Brillouin-amplified wave $E_-(x)\, e^{-ikx}$ plays

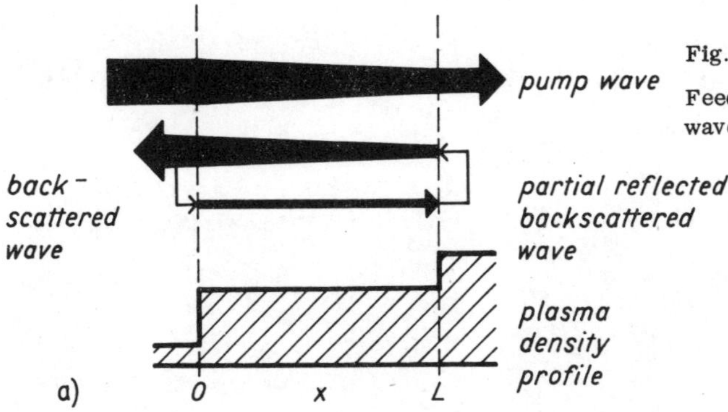

Fig. 9.18 (a)

Feedback scheme of the scattered wave at SBS

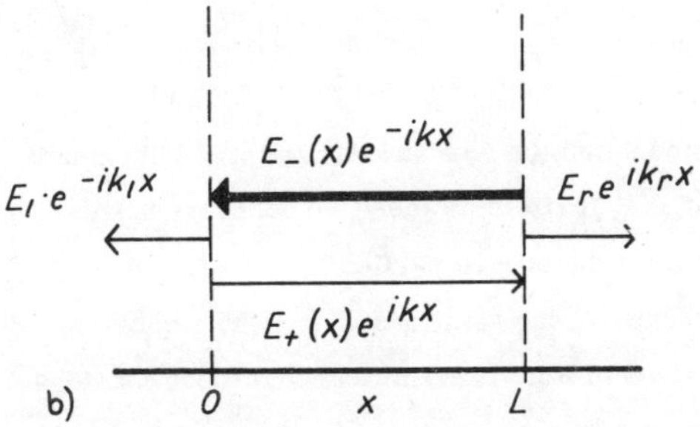

Fig. 9.18 (b)

Scheme for the calculation of the feedback factor. All of the waves shown have the frequency of the backscattered wave. $E_-(x)\,e^{-ikx}$ plays the role of the driver wave. k, k_l, k_r are the wave numbers of the plasma and the adjacent regions, respectively

the role of a driver wave. The jump conditions at the edges of the plasma resonator are

$$E_l = E_+ + E_-(0)\,, \quad -k_l E_l = kE_+ - kE_-(0)\,,$$

$$E_+ e^{ikL} + E_-(L)\,e^{-ikL} = E_r\,, \quad kE_+ e^{ikL} - kE_-(L)\,e^{-ikL} = k_r E_r$$

where k, k_r, k_l are the wave numbers in the plasma and the adjacent regions, respectively. From these conditions the equation

$$E_-(L) = R_r R_l\, e^{2ikL}\, E_-(0) \tag{9.92}$$

can easily be derived which represents the desired feedback relation. R_l, R_r are the reflectivities of the resonator boundaries, given by

$$R_l = \frac{k - k_l}{k + k_l}\,, \quad R_r = \frac{k - k_r}{k + k_r}\,. \tag{9.93}$$

Following Eq. (9.92), the feedback factor can be expressed as

$$f = R_l R_r e^{2ikL} , \qquad (9.94)$$

being in general a complex quantity and exhibiting a strong dependence on the length L of the scattering box.

The mode coupling Eq. (9.38) - (9.40) now have to be solved under the boundary conditions

$$E_1(0, t) = E_o , \quad q(0, t) = 0 \qquad (9.95)$$

$$E_2(L, t) = f E_2(0, t) . \qquad (9.96)$$

Parametric theory: We start again by looking for solutions within the framework of the parametric theory. Setting $E_1(x, t) = E_o$, using the ansatz E_2^*, $q \propto \exp i(Kx - \Omega t)$ and performing the same calculations as in Section 9.2 we arrive with the help of the results of the Appendix at an eigenvalue equation

$$\cos\frac{\pi}{2}P - i\frac{F + iS}{P}\sin\frac{\pi}{2}P = f e^{-i\frac{\pi}{2}(F + iS)} \qquad (9.97)$$

similar to Eq. (9.45) with an additional term that contains the feedback strength f. This condition has to be satisfied by the generalized frequency F in order to meet the boundary conditions (9.95), (9.96). The quantities F, S, P have been introduced earlier in Eqs. (9.46) - (9.49). As a consequence of a finite feedback strength f , the solutions F = F(Y) remain no longer purely imaginary. Fig. 9.19 displays solutions of Eq. (9.97) for three feedback factors $f = 0.01$ (a), $f = -0.01$ (b) and $f = 0.01\,i$ (c). Real and imaginary part of $Z = F + iS$ are plotted versus Y. The topology of the solutions has significantly changed compared with the feedback-free curves shown in Fig. 9.8. It is important to note, however, that the appearance of a finite $\mathrm{Re}(F) \propto \mathrm{Re}(\Omega)$ does not automatically result in oscillatory behaviour of the scattered intensity. This can simply be seen by going back to the ansatz

$$E_2(x, t) \propto e^{-i\Omega t}(Ae^{iK_1x} + Be^{iK_2x})$$

where $K_{1,2} = K_{1,2}(\Omega)$. When assuming that only a single mode is excited we find the scattered intensity I(t) to vary as

$$I(t) = |E_2(0, t)|^2 \propto |A + B|^2 e^{2\Gamma t}$$

($\Gamma = \mathrm{Im}\,\Omega$), thus showing no scillations. In the case when two modes are excited (Fig. 9.19 demonstrates that more than one solution may exist to a given Y), we have

$$I(t) = \left| C_1 e^{-i\mathrm{Re}\Omega_1 t} + C_2 e^{-i\mathrm{Re}\Omega_2 t} e^{(\Gamma_2 - \Gamma_1) t} \right|^2 \cdot e^{2\Gamma_1 t}$$

where $\Omega_1 \neq \Omega_2$, $\Gamma_1 > \Gamma_2$ and $C_{1,2}$ are the weigths of the modes.

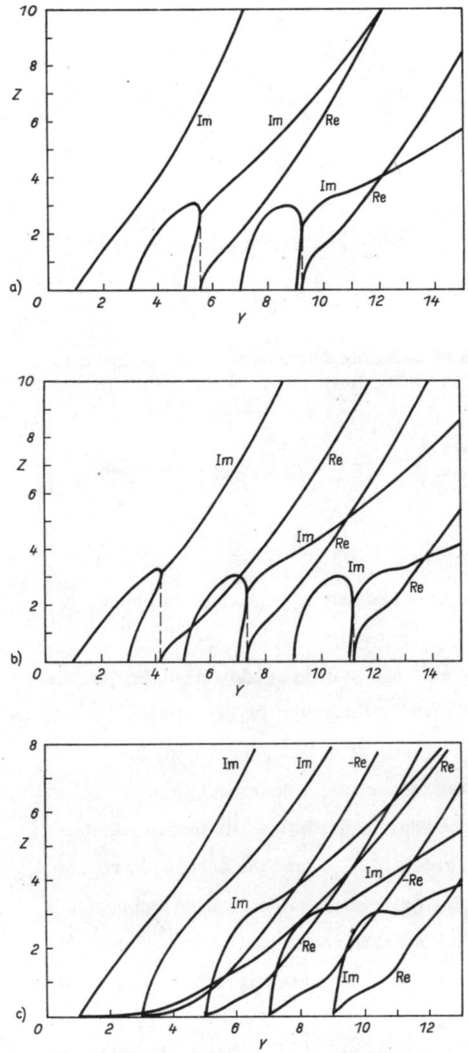

Fig. 9.19 Solutions of Eq. (9.97) describing SBS under the influence of a feedback of the scattered wave within the parametric theory; $Z = F + iS$ versus Y for three feedback factors, $f = 0.01$ (a), $f = -0.01$ (b), $f = 0.01\,i$ (c)

Due to $\Gamma_1 > \Gamma_2$ the first term dominates the solution with growing time, that is, only transient oscillations can be expected.

The only exception is the case of a real feedback factor. In this case two branches of solutions coalesce such that their imaginary parts coincide while their real parts differ in sign (see Figs. 9.19 (a), (b)). The scattered intensity then varies as

$$I(t) \propto \left[|C_1|^2 + |C_2|^2 + 2\,\mathrm{Re}\left\{ C_1 C_2^{*} e^{-2i\mathrm{Re}\Omega t} \right\} \right] e^{2\Gamma t}$$

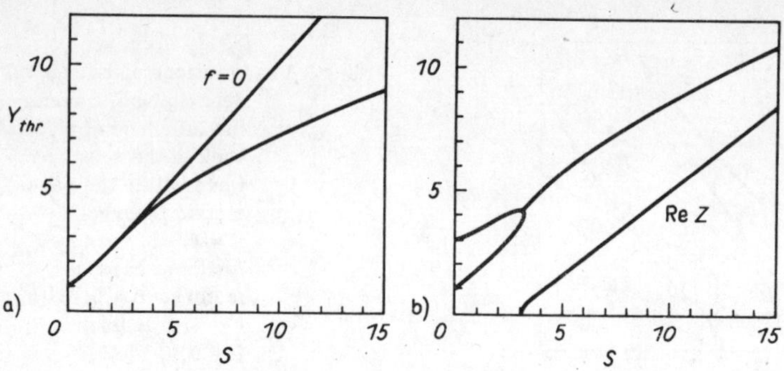

Fig. 9.20 Lowest SBS threshold versus generalized damping decrement S for feedback factors f = 0.01 (a) and f = - 0.01 (b). In (a) the feedback-free plot and in (b) Re(Z) at threshold are shown in addition

and exhibits continual oscillations. For equal weigths $C_1 = C_2 = C$ we get particularly

$$I(t) \propto 2\,|C|^2 (1 + 2\cos 2\mathrm{Re}\,\Omega\, t)\, e^{2\Gamma t} .$$

In this case the growth of the scattered intensity is accompanied by pulsations in which I (t) periodically passes through zero.

We now look for the lowest instability threshold as function of the generalized damping decrement S in the case of a real feedback factor. Following the graph Im (Z) versus Y in Fig. 9.19 in direction of growing Y, the threshold Y_{thr} is defined as that point where Im(F) $\propto$ Im(Ω) = 0, that is, Im(Z) = S. Obviously, for zero ion acoustic damping this condition is fulfilled at the point where the graph joins the Y-axis while a finite S shifts the threshold toward higher values. Fig. 9.20 illustrates how the threshold Y_{thr} and the associated real part Re(F) at threshold vary with the generalized damping decrement S for positive and negative feedback factors. (f = 0.01 (a), f = - 0.01 (b)). Note that the threshold decreases continuously relative to the feedback-free limit with increasing S. The most interesting feature, however, is that a real part of F (that is, of Ω) in the case of a negative feedback strength appears only when S exceeds a critical value S^*. Fig. 9.20 (b) indicates a linear increase of Re(F) with S, following a relation of the form

$$\mathrm{Re}(F) = \mathrm{Re}(Z) \simeq C(S - S^*), \qquad S > S^* . \tag{9.98}$$

Thereby the quantities C, S^* depend roughly logarithmically on f and can be fitted within 10 % by the formulas

$$S^* = \ln(1/|f|), \qquad C = (2/S^*)^{1/2} . \tag{9.99}$$

Similarly, a linear variation of Re(F) with S is obtained also for other feedback strengths as shown in Fig. 9.21.

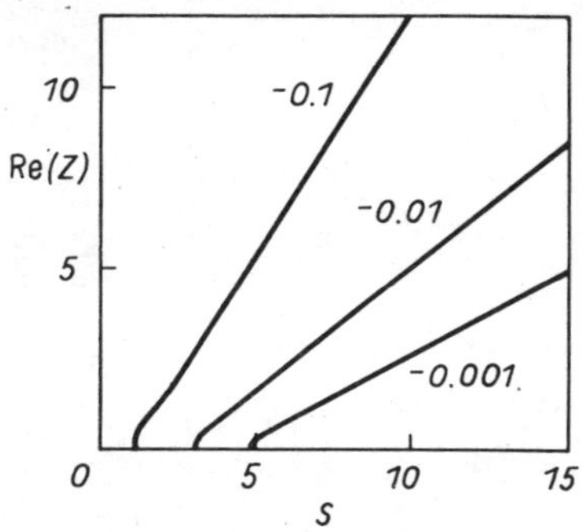

Fig. 9.21 Re Z at threshold as function of the generalized damping decrement S for three negative feedback strengths f (curve parameter)

Mode coupling theory :

The results of the above parametric approximation have been tested by the mode coupling theory. Numerical solutions of Eqs. (9.38) - (9.40) under the boundary conditions (9.95), (9.96) for a great variety of parameters Q, S, f revealed that the behaviour of the nonlinear system with respect to the oscillation period is very well described by the predictions of parametric theory at the instability threshold:

(a) The necessary condition for continued pulsations to occur is $f < 0$, $S > S^*$;

(b) the oscillation period is in accordance with Eq. (9.98) and depends only weakly on Y;

(c) only transient pulsations occur for complex feedback strength; no pulsations appear for $f > 0$.

Following Eq. (9.98) we find that the scaling of the oscillation period $T = \pi/\Omega$ is approximately given by

$$T = 2\sqrt{\varepsilon}\,\frac{L}{\lambda_0}\,\frac{1}{C(S - S^*)}\,T_s \tag{9.100}$$

where $T_s = 2\pi/\omega_s$ is the ion acoustic period. In the limit $S \gg S^*$ (for $|f| \geq 10^{-4}$ equivalent to $\varepsilon^{1/2} L/\lambda_0\ \sigma/\omega_s \gg 1$) the scaling becomes particularly simple. Eq. (9.100) then reduces to

$$T \simeq \frac{\pi}{C\sigma} = \left(\frac{\ln(1/|f|)}{2}\right)^{1/2} \frac{1}{\sigma}. \tag{9.101}$$

That is, for strong ion acoustic damping and/or long plasmas the oscillation period is inversely proportional to the ion acoustic damping rate and the pulsations may be called relaxation determined.

Fig. 9.22 shows, as an example, a typical solution of the mode coupling equations for $f = -0.01$. The time history of the relative backscattered intensity $|R(t)|^2$ is plotted for a pump amplitude $E_0 = 0.13$. The plot indicates only a slight decrease of period T with increasing pump amplitude. The time averaged SBS reflectivity is fairly reduced relative to the feedback-free case. In Fig. 9.22 (b), (c) spatial amplitude profiles of the two decay waves are plotted at two times, where $|R|^2$ passes through a maximum (a) and through zero (b). Fig. 9.22 (c) reveals that zero scattering is produced via coherent distruction according to the relation (9.67)

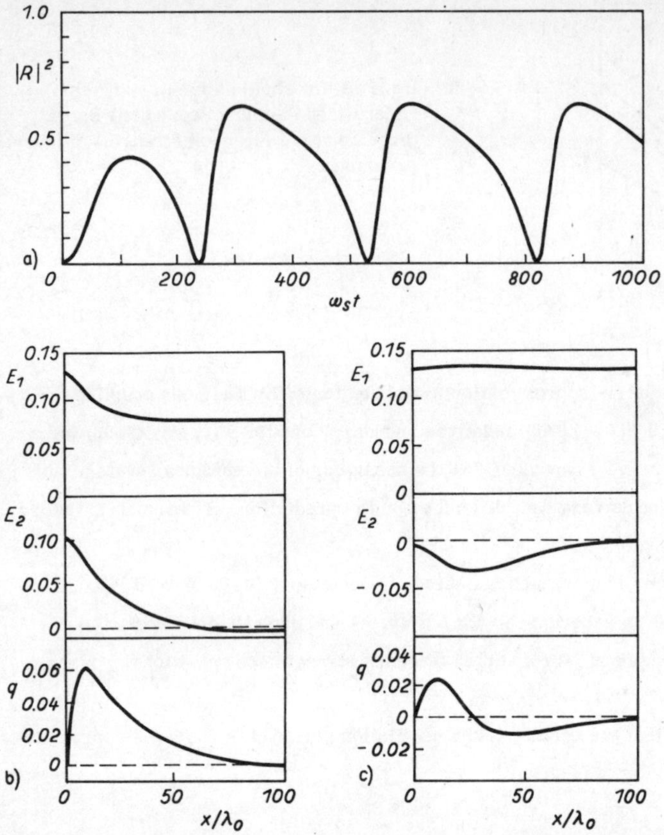

Fig. 9.22 SBS reflectivity versus time (a) and spatial amplitude profiles at maximum reflectivity (b) and at zero reflectivity (c) in the presence of a feedback of the backscattered wave. Parameters: $E_o = 0.13$, $N_o/N_c = 0.3$, $L/\lambda_o = 100$, $\sigma/\omega_s = 0.025$, $f = -0.01$ ($Y \simeq 10$, $S \simeq 8.4$)

$$R(t) = \tanh \frac{N_o}{N_c} \frac{1}{4\sqrt{\varepsilon}\, c} \int_0^L q(x, t)\, dx$$

which follows from Eqs. (9.38), (9.39) neglecting time derivatives.

The fact that sufficient ion acoustic damping is necessary for the feedback to produce pulsations of the scattered intensity can be understood as follows: the amplitude of the right-travelling ion acoustic wave has to be small enough at $x = L$ in order to allow the feedback-produced density disturbance to initiate the development of a backward-sweeping region with phase-inverted scattering. For a too low ion acoustic damping a phase-inverted field amplitude $E_2(L, t)$ actually continues to exist but the density disturbance at $x = L$ turns to be dominated by the right-travelling ion acoustic waves, which stops the pulsations.

The pulsations described above differ in nature from those found in certain optical bi-stable systems. The latter ones, described for example in /111 - 113/ are caused by multivalued solutions for the stationary reflectivity/transmission as function of the incident energy. In contrast, the stationary mode coupling equations in our case have no solutions for parameter combinations which result in pulsating solutions of the time-dependent equations.

9.6 Strong-coupling effects

Strong coupling in the sense introduced in Section 9.1 implies a description of the ion acoustic disturbance by the full wave equation. That is, variations on the ion acoustic time scale $1/\omega_s$ have to be covered by the slowly varying amplitudes and the process cannot longer be considered as a three-wave interaction. While the equations for the pump wave and the scattered wave remain unchanged, the ion acoustic amplitude q (x, t) in the frame of strong coupling is governed by the equation

$$\frac{\partial^2 q}{\partial t^2} + 2\sigma \frac{\partial q}{\partial t} - 2i\,\omega_s c_s \frac{\partial q}{\partial x} + \omega_s^2 q = -\frac{1}{2}\omega_s^2 E_1 E_2^* . \tag{9.102}$$

It is capable of describing general density perturbations exhibiting a spatial variation $\propto \exp(ik_s x)$ including ion acoustic waves in both directions as well as standing density structures.

While a non-zero initial amplitude of the ion acoustic wave is necessary to initiate the SBS instability within the three-wave interaction model, amplification occurs without such an initial level in the presence of a partial reflecting boundary as a consequence of strong coupling effects. To illustrate this we consider the initial phase of SBS when a partial reflecting boundary (reflectivity R_w) is present. Ignoring damping the early stage of the ion acoustic disturbance is described by the equation

$$\frac{\partial^2 q}{\partial t^2} + \omega_s^2 q = -\frac{1}{2}\omega_s^2 R_w E_o^2$$

which follows immediately from Eq. (9.102). The source term represents the ponderomotive force resulting from the superposition of ingoing and partially reflected pump wave. The solution satisfying the initial conditions $q = 0$, $\partial q/\partial t = 0$ at $t = 0$ is

$$q = \frac{1}{4} R_w E_o^2 \left(e^{-i\omega_s t} + e^{i\omega_s t} - 2\right) .$$

It can be interpreted as superposition of two counter-propagating acoustic waves and a standing density grating. Hence, the presence of a partial reflecting layer creates a right-travelling ion wave at the very beginning and thus initiates the SBS instability. Assuming an energy reflection of 1 % ($R_w = 0.1$) a pump wave with amplitude $E_o = 0.1$ causes an initial level $q_o = 2.5\ 10^{-3}$.

The subsequent time development of the SBS reflectivity as a result of a strong-coupling calculation is illustrated in Fig. 9.23. The choice of parameters is the same as in Fig. 9.10. As a

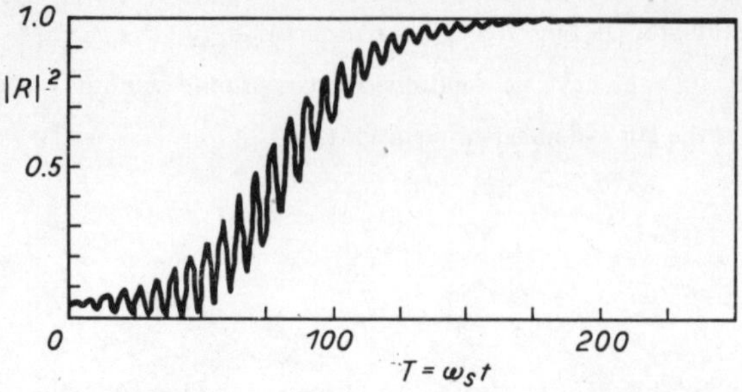

Fig. 9.23 SBS reflectivity versus time in the presence of a partially reflecting right-hand boundary of the scattering box in strong coupling description; $E_o = 0.2$, $N_o/N_c = 0.3$, $L/\lambda_o = 25$, no ion acoustic damping, reflectivity of the boundary $R_w = 0.2$

characteristic feature, the scattered intensity oscillates with the ion acoustic frequency, accompanied by phase oscillations of the same frequency. These oscillations are the result of the superposition of the SBS amplified wave with frequency $\omega_o - \omega_s$ and the partially reflected pump wave with frequency ω_o. The modulation decreases with growing $|R|^2$ and vanishes for $|R|^2 \rightarrow 1$. Obviously, the reflecting layer loses influence on the scattering process when the pump wave energy is backscattered before reaching it.

As an interesting effect, at strong pump amplitudes and sufficiently high ion acoustic damping backscatter pulsations occur which seem to be closely related to the relaxation determined feedback oscillations studied in the preceding section. While the pulsations within the weak-coupling model could be explained as an effect of a feedback due to double reflection of the Brillouin-amplified wave at the edges of a spatially limited scattering box, the pulsations here occur even without a second partially reflecting boundary. The effects is demonstrated in Fig. 9.24. It shows the time history of $|R|^2$ for two damping decrements differing by a factor two. From this figure it is seen that besides weak oscillations with the ion acoustic frequency there are pulsations with a longer period that turn out to be inversely proportional to the damping decrement σ. This indicates the presence of a self-induced negative feedback; probably either by interac-

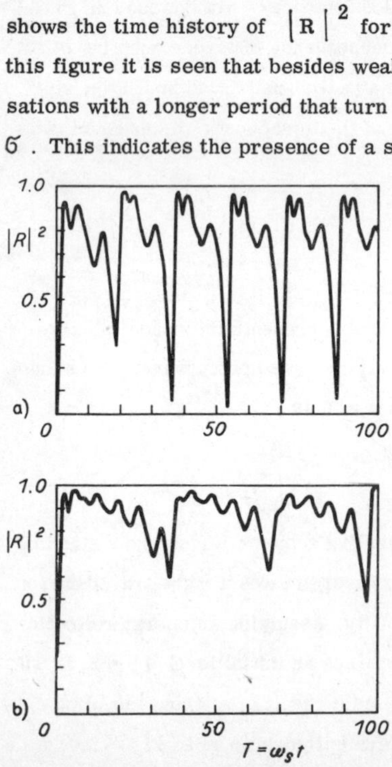

Fig. 9.24 The same as Fig. 9.23 in the case of strong pump intensity and strong ion acoustic damping; $N_o/N_c = 0.3$, $L/\lambda_o = 50$, $\sigma/\omega_s = 0.25$ (a), $\sigma/\omega_s = 0.125$ (b), $E_o = 0.75$

tion of the partially reflected pump wave with left-travelling ion acoustic waves or by continual reflection of the scattered wave at zero-frequency density perturbations. In contrast to the pulsations originating from the discrete feedback the effect does not sensitively depend on the scattering length L and proves to be independent of the sign of R_w.

The suggestion of a negative feedback is supported by the plots in Fig. 9.25 which show spatial amplitude profiles of $|q|$ and $|E_2|$. These plots are similar to those in Fig. 9.22. In particular, the phase jump of q in Fig. 9.25 (a) indicates that minimum scattering is again produced by coherent distruction of the scattered wave according to Eq. (9.67). When comparing the spatial profiles one should take into consideration, however, that in Fig. 9.25 (a), (b) the amplitudes E_1, E_2 contain contributions of both the scattered frequency and the pump frequency.

In this context we point to recently observed 100 % ion density modulations at $k = k_s$ with a period of about 100 ps (corresponding to several ion acoustic periods) in an SBS experiment with a CO_2-laser, with use of Thomson scattering /20/. Although there was no possibility to observe corresponding temporal modulations in the backscattered radiation, the experimental results might be interpreted in terms of feedback pulsations. This is suggested by the following arguments: (a) The observed period is in reasonable agreement with the prediction of Eq. (9.101). Assuming ion Landau damping, one gets $\sigma / \omega_s \lesssim 0.1$ for $T_e \simeq 5T_i$. If we assume further a feedback strength of $f = -10^{-4}$, Eq. (9.101) gives $\omega_s T \simeq 50$, equivalent to several ion acoustic periods. (b) The pulsation period decreases with increasing pump power. This can be understood in terms of the above model as follows: If we accept that the self-induced feedback increases with increasing pump power, then the absolute value $|f|$ of the feedback factor increases, too. Following Eq. (9.101), this results in a reduced pulsation period. A nonlinear ion acoustic damping mechanism would act in the same direction. (c) The observed pulsations of the Thomson-scattered light need not be assumed to be produced by a spatially homo-

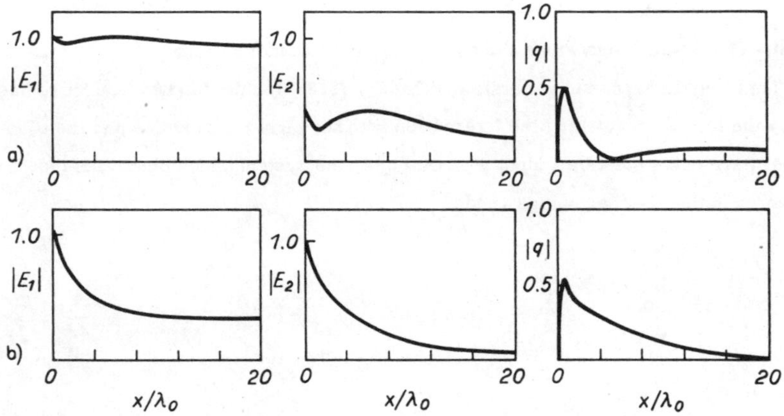

Fig. 9.25 Spatial amplitude profiles (associated with Fig. 9.24 (a)) at minimum reflectivity (a) and maximum reflectivity (b)

geneous time-varying ion acoustic amplitude, which is rather unlikely, but may be achieved by phase-mixing effects in a way similar to that which brings about the pulsations of $|R|^2$ Obviously, any time modulation of the intensity of the scattered wave is transformed to the transmitted pump energy. Hence, when the Brillouin-active region extends in space to a plasma exceeding the critical density, the effects occurring at the critical layer, such as harmonic generation, suffer from an oscillating pump intensity and the same pulsations might be found also in the harmonic radiation.

9.7 SBS saturation by ion acoustic nonlinearities

In the preceding sections the treatment of stimulated Mandelstam-Brillouin scattering was based on a linear wave equation for the ion acoustic wave. This is justified as long as the relative density disturbance $|\delta N|/N_o$ remains small against one. Calculations on the space-time development of the SBS instability revealed, however, that this assumption is violated even for moderate pump amplitudes. From Fig. 9.10 it is seen, for example, that a density disturbance of the order $|\delta N|/N_o \simeq 0.4$ is generated by a pump wave with amplitude $E_o = 0.2$ in a plasma with $N_o = 0.3\ N_c$. To produce the same relative density perturbation in a plasma with $N_o = 0.1\ N_c$, a pump power of $E_o = 0.1$ is already sufficient. This makes a linear treatment of the ion acoustic wave questionable.

The most elementary nonlinear phenomenon at the propagation of a monochromatic ion acoustic wave is the generation of harmonics (see Section 8.1). This process distributes the energy of the resonantly enhanced wave with $k_s = 2k_o$ over modes with $2k_s$, $3k_s$ and thus hampers the growth of the basic wave. At sufficiently high amplitudes and/or long scattering regions eventually wave breaking occurs; both effects represent possible saturation mechanisms of the SBS instability /81 - 88/.

Saturation by harmonic generation

The system (9.38) - (9.40) which describes the space-time evolution of SBS can simply be extended to include ion acoustic harmonic generation. The Eq. (9.38) for the fundamental wave has to be completed by the terms describing the interaction with the higher harmonics and the equations for the higher harmonics must be added. For example, the system which describes the stationary problem including five harmonics reads

$$\frac{dE_+}{dx} = -\frac{\omega}{4\sqrt{\varepsilon}\,c}\,\frac{N_o}{N_c}\,q_1 E_- ,$$

$$\frac{dE_-}{dx} = -\frac{\omega}{4\sqrt{\varepsilon}\,c}\,\frac{N_o}{N_c}\,q_1^{*} E_+ ,$$

$$\frac{dq_1}{dx} = -\frac{\sigma_1}{c_s}\,q_1 + \frac{k_s}{4}\,E_+E_-^{*} + k_s\,(-q_2 q_1^{*} - q_3 q_2^{*} - q_4 q_3^{*} - q_5 q_4^{*})$$

$$\frac{dq_2}{dx} = -\frac{\sigma_2}{c_s} q_2 + 2k_s\left(-q_3 q_1^* - q_4 q_2^* - q_5 q_3^* + \frac{1}{2} q_1^2\right),$$

$$\frac{dq_3}{dx} = -\frac{\sigma_3}{c_s} q_3 + 3k_s\left(-q_4 q_1^* - q_5 q_2^* + q_1 q_2\right),$$

$$\frac{dq_4}{dx} = -\frac{\sigma_4}{c_s} q_5 + 5k_s\left(q_1 q_4 + q_2 q_3\right). \tag{9.103}$$

where E_+, E_- denotes pump wave and scattered wave, respectively. σ_1, σ_2, ... are the damping decrements of the harmonics. The harmonics are not involved in the SBS mechanism, they grow only due to their coupling to the fundamental wave q_1.

Fig. 9.26 shows as a result of a numerical solution of system (9.103) the reflectivity as function of the generalized pump amplitude Y for different numbers of harmonics included (damping ignored). Slightly above the threshold $Y = 1$ the reflectivity is seen to be strongly reduced no matter how many harmonics are taken into consideration. With increasing Y, however, the graphs corresponding to different harmonic number diverge. Particularly the graph $N = 2$ is striking in that it shows that the instability cannot develop at all in this case. The reason for this singular behaviour is the peculiarity of the energy exchange between the two waves: when interacting only with its second harmonic the fundamental wave can lose all its energy in favour of the harmonic. This behaviour can simply be demonstrated by solving the pair of equations

$$\frac{dq_1}{dx} = -q_1 q_2, \qquad \frac{dq_2}{dx} = q_1^2$$

which describe the interaction of the fundamental wave and the second harmonic. The solution satisfying the boundary conditions $q_1(0) = q_{10} = 0$, $q_2(0) = 0$ is

$$q_1(x) = q_{10}\,\mathrm{sech}(q_{10}x), \qquad q_2(x) = q_{10}\tanh(q_{10}x).$$

That is, the energy of the basic wave is completely converted into the second harmonic for $x \to \infty$.

Note that even for $N = 5$ the instability rapidly saturates at reflectivities close to one for

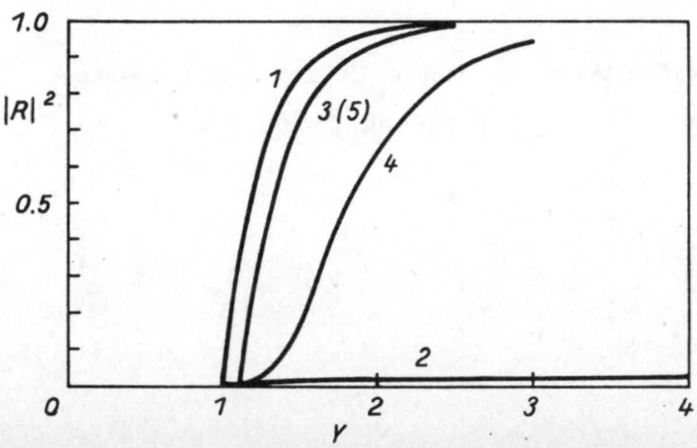

Fig. 9.26 Steady-state SBS reflectivity versus generalized pump amplitude Y for different numbers of ion acoustic harmonics included (curve parameter). The graph $N = 5$ is identical with that for $N = 3$ within the graphical accuracy

growing Y. This indicates that ion acoustic harmonic generation cannot be considered as an effective saturation mechanism.

Saturation by wave breaking

Keeping the number of the excited harmonics at a certain limit prevents an initially sinusoidal wave from breaking as it propagates through a weakly nonlinear medium. Wave breaking, however, is accompanied by an irreversible loss of wave energy as shown in Section 8.1 and thus seems to be a better candidate for SBS saturation as the excitation of a limited number of ion acoustic harmonics. We shall see that the nonlinear behaviour of the ion acoustic wave can be described by a simple quadratic damping term in the wave equation. This term approximately accounts for both effects, harmonic generation and wave breaking. It simplifies the analysis considerably because additional equations for the harmonics are no longer necessary.

In Section 8.1 it was shown that the time evolution of an initially sinusoidal wave profile proceeds in two stages: (a) steepening towards a saw-tooth profile ($0 \lesssim t \lesssim t_c$), (b) amplitude relaxation of the saw-tooth profile ($t \gtrsim t_c$). t_c is given by the conditions (8.12). The Fourier spectrum of the initial profile is simply

$$(q_1, q_2, \ldots)/_{t=0} = \frac{q_o}{2}(1, 0, 0, \ldots) .$$

The saw-tooth profile (8.15) has the spectrum

$$(q_1, q_2, \ldots)/_{t=t_c} = \frac{q_o}{2}(1, 1/2, 1/3, \ldots) . \tag{9.104}$$

Comparing both spectra we find that the fundamental amplitude has reduced by a factor $2/\pi$ during the time interval $0 \lesssim t \lesssim t_c$. The subsequent time development of the saw-tooth profile is characterized by a relaxation of all Fourier components according to $1/(t + (t - t_c)/\pi t_c)$ with maintaining the amplitude ratios fixed by Eq. (9.104). Especially we have for the fundamental wave

$$q_1(t) = \frac{q_1(t_c)}{1 + \frac{t - t_c}{\pi t_c}} = \frac{2}{\pi} q_o \frac{1}{1 + \frac{t - t_c}{\pi t_c}} \tag{9.105}$$

($t_c \lesssim t$). Extrapolating this relation back to $t = 0$ we find

$$q_1(0) \simeq 0.93\, q_o$$

That is, Eq. (9.105) approximately describes the behaviour of $q_1(t)$ also at the initial stage where no analytical expression is available. Time derivation of Eq. (9.105) gives

$$\frac{dq_1}{dt} = -\frac{2}{\pi^2} q_o^2 \frac{1}{\left(1 + \frac{t - t_c}{\pi t_c}\right)^2} = -\frac{1}{2}\omega q_1^2 \tag{9.106}$$

where the relation $t_c = 1/\omega q_o$ has been used.

The spatial analog to Eq. (9.106) is

$$\frac{dq_1}{dx} = -\frac{1}{2} k\, q_1^2 . \tag{9.107}$$

Including this damping decrement in the mode coupling equations for the stationary problem we have the system

$$\frac{dE_1}{dx} = -\frac{\omega_1}{4\sqrt{\varepsilon}\, c} \frac{N_o}{N_c} q\, E_2 ,$$

$$\frac{dE_2}{dx} = -\frac{\omega_1}{4\sqrt{\varepsilon}\, c} \frac{N_o}{N_c} q\, E_1 ,$$

$$\frac{dq}{dx} = -\frac{\sigma}{c_s} q - \frac{1}{2} k_s^2 q^2 - \frac{k_s}{4} E_1 E_2 . \tag{9.108}$$

where all quantities are taken as real and $q > 0$. Results of an integration of this system are presented in Fig. 9.27. The plots show the reflectivity versus Y for two plasma densities. The nonlinear damping term is seen to cause a significant reduction of the saturated SBS reflectivity. Thus ion acoustic wave breaking is suggested to act as effective saturation mechanism for the SBS instability.

Similar dependencies are calculated in /86/, also based on a quadratic damping term in the ion acoustic wave equation.

The results displayed in Fig. 9.27 are supported by calculations within the framework of the fully nonlinear description of the ion dynamics (see Chapter 11).

9.8 The influence of plasma flow on SBS

In the previous sections pertaining to SBS we considered the Brillouin-active medium to be at rest in the laboratory frame. At laser target interaction, for example, this assumption is no longer valid. The plasma produced at the target expands radially forming a medium with inhomogeneous profiles of density and velocity. In the underdense outer parts of the plasma the flow velocity may exceed the ion acoustic speed.

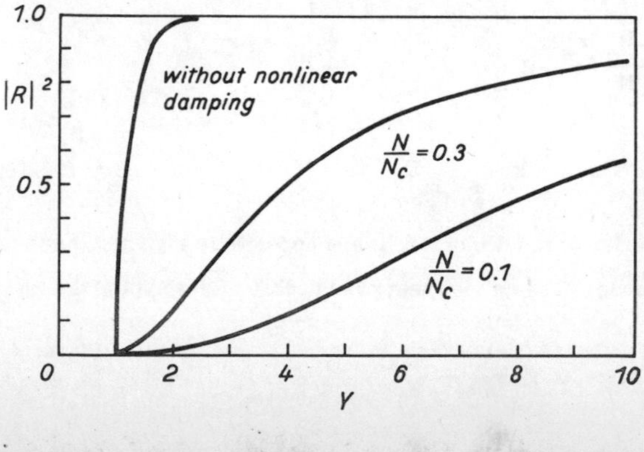

Fig. 9.27 The same as Fig. 9.26, with a nonlinear ion acoustic damping term included which accounts approximately for wave braking; no linear ion acoustic damping

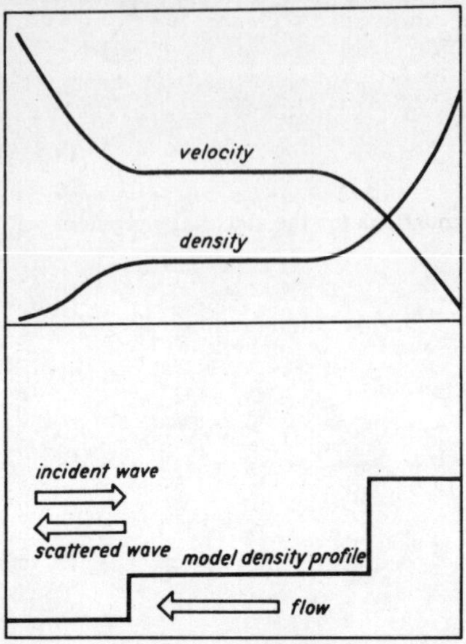

Fig. 9.28 Plasma model used for the study of SBS under the influence of a plasma flow

In order to reduce the analysis, we examine here the space-time evolution of SBS in a strongly simplified step-like plasma profile as sketched in Fig. 9.28. This is suggested by calculations on the plasma dynamics under the action of a strong incident electromagnetic wave in the presence of plasma flow carried out in Chapters 10 and 11. Fig. 10.11, for example, demonstrates that the combined action of ponderomotive force and plasma flow tends to form an underdense plasma plateau with supersonic flow, adjacent to an overdense region.

The presence of a non-zero plasma flow creates additional terms in the ion acoustic wave equation. In the one-dimensional version (flow velocity v_o parallel to the x-axis) we have

$$\frac{\partial^2 \delta N}{\partial t^2} + \sigma\left(v_o \frac{\partial \delta N}{\partial x} + \frac{\partial \delta N}{\partial t}\right) + 2v_o \frac{\partial^2 \delta N}{\partial x \partial t} - (c_s^2 - v_o^2)\frac{\partial^2 \delta N}{\partial x^2} = \frac{1}{4} N_o c_s^2 \frac{\partial^2}{\partial x^2} |E|^2 . \tag{9.109}$$

The only modification that arises in the mode coupling equations is the appearance of the new group velocity $v_o + c_s$ of the ion acoustic waves instead of c_s. Thus, the system now reads in its simplest form

$$\frac{\partial E_1}{\partial t} + \sqrt{\varepsilon}\, c \frac{\partial E_1}{\partial x} = -\frac{\omega}{4}\frac{N_o}{N_c} q\, E_2 ,$$

$$\frac{\partial E_2}{\partial t} - \sqrt{\varepsilon}\, c \frac{\partial E_2}{\partial x} = \frac{\omega}{4}\frac{N_o}{N_c} q^* E_1 ,$$

$$\frac{\partial q}{\partial t} + (c_s + v_c)\frac{\partial q}{\partial x} = -\sigma q + \frac{1}{4} k_s c_s E_1 E_2^* . \tag{9.110}$$

As long as $v_o > -c_s$ the group velocities of the decay waves are opposite in sign and the instability remains absolute in nature. Hence, parametric theory is capable of describing the influence of plasma flow on the threshold as well as the growth rate in this case. The theory of Section 9.2 is applicable; thresholds and growth rates can be expressed in terms of the quantities F, S, Y defined in Eqs. (9.46) - (9.48), modified by the finite Mach number $M = v_o/c_s$ as follows:

$$F = 4\sqrt{\varepsilon}\,\frac{L}{\lambda_o}\,\frac{1}{1+M}\,\frac{\Omega}{\omega_s},$$

$$S = 4\sqrt{\varepsilon}\,\frac{L}{\lambda_o}\,\frac{1}{1+M}\,\frac{\sigma}{\omega_s},$$

$$Y = \left(2\,N_o/N_c\,\frac{1}{1+M}\right)^{1/2}\,\frac{L}{\lambda_o}\,E_o\,. \tag{9.111}$$

In absence of damping the threshold $E_{o,\,thr}$ is again given by the condition $Y = 1$, i. e.,

$$\left(2\,N_o/N_c\,\frac{1}{1+M}\right)^{1/2}\,\frac{L}{\lambda_o}\,E_{o,\,thr} = 1\,. \tag{9.112}$$

For $M > 0$, i. e., when the plasma flow has the direction of the pump wave, the threshold is increased. In the opposite case $M < 0$ we get a lower threshold. The reduction of the thresholds in the latter case can be understood as an effect of a prolonged presence of the ion acoustic disturbances within the operation region of the instability owing to the reduced group velocity. In the singular case $v_o = -c_s$ $(M = -1)$ the group velocity is zero, i. e., the ion acoustic disturbances would persist and grow in time without bound. This particular case is interesting in connection with a non-uniform flow profile where locally the condition $v_o(x) = c_s$ is met. The neighborhood of this point proces to be a main source of scattered radiation, as will be seen in Chapter 11.

When the flow velocity against the direction of the pump wave propagation exceeds the ion acoustic speed, the decay waves propagate parallel and the instability changes its character to become convective. The boundary condition $q(0, t) = q_o$ then loses its meaning and has to be replaced by

$$q(L, t) = q_L \neq 0\,.$$

A non-zero steady-state scattering level can develop in this case only when a finite ion acoustic source $q_L \neq 0$ is present at the right-hand boundary.

Fig. 9.29 illustrates the evolution of stimulated Brillouin scattering in space and time in a flowing plasma with $M = -2$. An initial level $q_o(x, 0) = 0.01$ has been assumed, the same value has been taken as boundary value $(q_L = 0.01)$. As may be seen from this figure, a significant stationary reflection level develops for this choice of parameters. If the boundary value q_L is set equal to zero for $t > 0$ in the same run, only a transient reflectivity is produced which subsequently drops to zero.

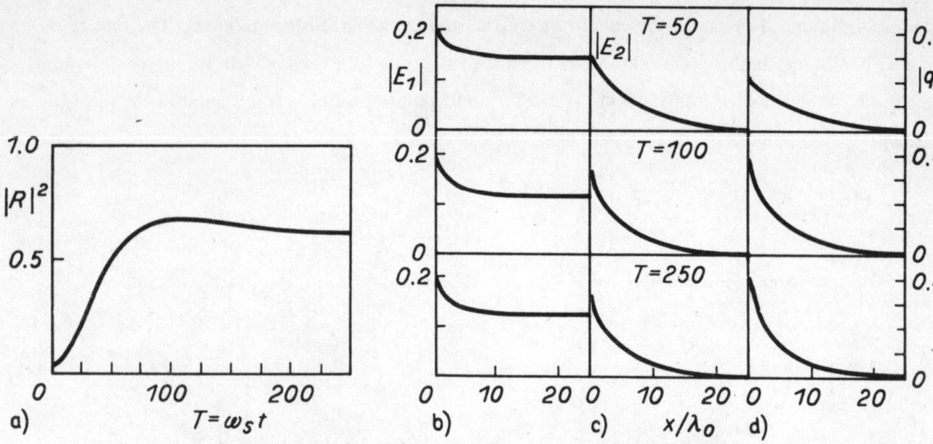

Fig. 9.29 Space-time development of a convective SBS instability in the presence of a supersonic plasma flow against the direction of the pump wave. Reflectivity versus time (a); spatial amplitude profiles of pump wave (b), scattered wave (c) and ion acoustic wave (d) at three times. $E_o = 0.2$, $N_o/N_c = 0.3$, $L/\lambda_o = 25$, $M = -2$, $q(x, 0) = q(L,t) = 0.01$, no ion acoustic damping

SBS feedback pulsations at supersonic plasma flow

The presence of a supersonic plasma flow against the pump wave direction offers the possibility of feedback effects on the scattering process also if ion acoustic damping is absent /74/. This is because the ion acoustic disturbance at the right-hand boundary is now determined by the noise level and thus can be assumed to be small independently of ion acoustic damping. A first approach to the behaviour of the scattering process under the influence of a feedback of the scattered wave is again provided by the parametric theory. The basic equations are

$$\frac{\partial E_2}{\partial t} - \sqrt{\varepsilon}\, c \frac{\partial E_2}{\partial x} = -\frac{\omega}{4} \frac{N_o}{N_c} q E_o , \tag{9.113}$$

$$\frac{\partial q}{\partial t} + (c_s + v_o) \frac{\partial q}{\partial x} = -\sigma q + k_s c_s E_2 E_o \tag{9.114}$$

with $E_1(x, t) = E_o = \text{const}$. Boundary conditions are

$$q(L, t) = q_o , \qquad E_2(L, t) = f E_2(0, t) . \tag{9.115}$$

Treating these equations in the standard way as described in the Appendix we find the equation

$$i \frac{(F + iS)}{P} \sin \frac{\pi}{2} P + \cos \frac{\pi}{2} P = \frac{1}{f} e^{i \frac{\pi}{2} (F + iS)} \tag{9.116}$$

as condition for the existence of a non-trivial solution. The quantities F, S, P, Y are now defined as follows:

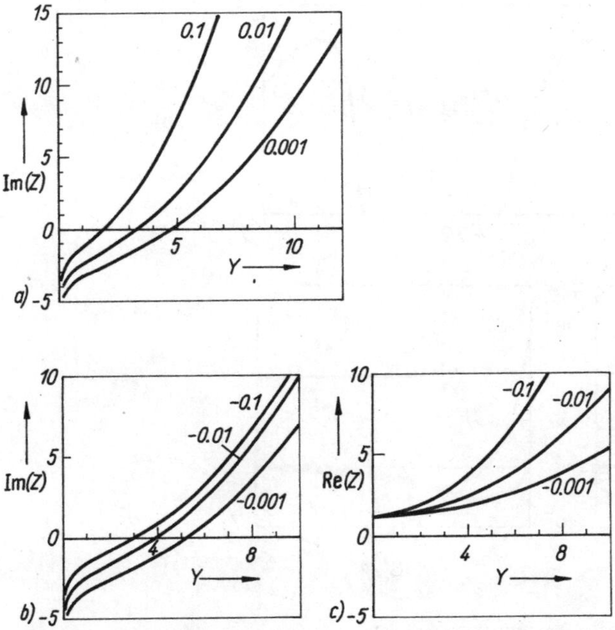

Fig. 9.30 Purely imaginary solutions of Eq. (9.116) in the case of positive feedback (a) and complex solutions (b), (c) for negative feedback; curve parameter: feedback strength f

$$Z = F + iS = \frac{4\sqrt{\mathcal{E}}}{|1 + M|} \frac{L}{\lambda_o} \frac{\Omega + i\sigma}{\omega_s} , \quad (9.117)$$

$$Y = \left(2 \frac{N_o}{N_c} \frac{1}{|1 + M|} \right)^{1/2} \frac{L}{\lambda_o} E_o , \quad (9.118)$$

$$P = (Z^2 - Y^2)^{1/2} , \quad (9.119)$$

with $M = v_o/c_s < -1$.

Fig. 9.30 exhibits solutions of Eq. (9.116) in the form $Z = F + iS$ versus the generalized pump amplitude Y. In case of a positive feedback strength (Fig. 9.30 (a)) there are only purely imaginary solutions. Similar to the situation of a plasma at rest (Section 9.5) a negative feedback produces complex solutions which give rise to the occurence of regenerative pulsations of the scattered energy. Figs. 9.30 (b), (c) show solutions for several negative feedback strengths.

Common to all cases is that the instability changes its nature to become absolute again as a consequence of the feedback. Obviously, the threshold depends on the feedback strength and goes to infinity as f goes to zero.

Numerical solutions of the mode coupling Eq. (9.110) under the boundary conditions (9.115) with $M < -1$ and negative feedback factors reveal that similarly the pulsation period $T = \pi/\Omega$

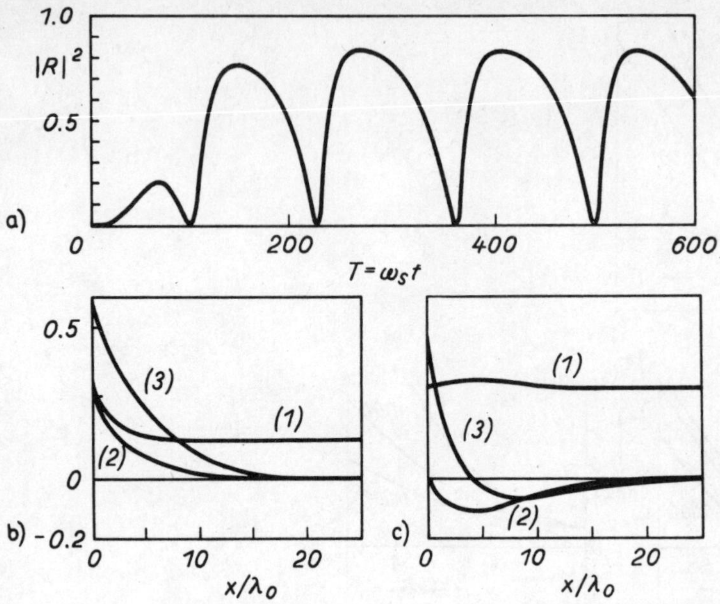

Fig. 9.31 Regenerative pulsations of the SBS reflectivity in the presence of both a supersonic flow against the pump wave and a feedback of the scattered wave. Reflectivity versus time (a); spatial amplitude profiles of pump wave (1), backscattered wave (2) and ion acoustic wave (3) for maximum reflectivity (b) and zero reflectivity (d). $E_o = 0.3$, $N_o/N_c = 0.3$, $L/\lambda_o = 25$, $M = -2$, $\sigma = 0$, $f = -0.01$

coincides with that predicted by the parametric theory at threshold. Fig. 9.31 shows as an example a solution for the following choice of parameters: $f = -0.01$, $E_o = 0.3$, $N_o/N_c = 0.3$, $M = -2$, $L/\lambda_o = 25$, $\sigma = 0$. There is only a weak dependence of T on the generalized amplitude Y. The period $\omega_s T \simeq 110$ in Fig. 9.31 corresponds to $F \simeq 2.5$.

From Fig. 9.30 it is seen that regardless of the feedback strength the real part of F at threshold is essentially constant ($\mathrm{Re}(F) \simeq 2.5$). This allows a simple scaling of the oscillation period. From

$$\mathrm{Re}(F) = \frac{4\sqrt{\mathcal{E}}}{|1+M|}\frac{L}{\lambda_o}\frac{\Omega}{\omega_s} \simeq 2.5$$

it follows

$$\omega_s T \simeq 5\sqrt{\mathcal{E}}\frac{L}{\lambda_o}\frac{1}{|1+M|} \tag{9.120}$$

or

$$T \simeq 0.8\sqrt{\mathcal{E}}\frac{1}{|1+M|}\frac{L}{\lambda_o}T_s . \tag{9.121}$$

where $T_s = 2\pi/\omega_s$ is the ion acoustic period.

Fig. 9.32 demonstrates the influence of ion acoustic damping on both threshold and pulsation period. Y_{thr} and Re(F) at threshold are plotted versus the generalized damping decrement S

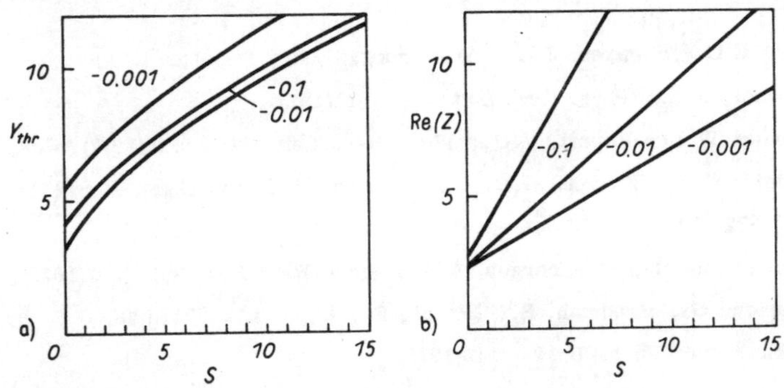

Fig. 9.32 Threshold Y_{thr} (a) and Re(Z) at threshold (b) as function of the generalized damping decrement S as predicted by parametric theory (Eq. 9.116)

for three negative feedback strengths. From Fig. 9.32 (b) it is seen that the graphs F (S) belonging to different feedback factors join the ordinate approximately at the same position $F \simeq 0.25$ for $S \longrightarrow 0$. This again demonstrates that the pulsation period does not depend on the feedback strength in absence of ion acoustic damping. Note that a strictly linear variation of F with S is reproduced, as already found in the case without plasma flow. The factor C in the corresponding relation

$$F = C\ S + 2.5$$

coincides with that in Eq. (9.98). Hence, in case of strong damping the same dependence $T \propto 1/\sigma$ of the pulsation period on the damping decrement is reproduced, as expressed by Eq. (9.101).

References

/1/ Goldman, L, Soures, J., Lubin, M.J.; Phys. Rev. Lett. 31, 1184(1973).

/2/ Basov, N.G., Krokhin, O.N., Pustovalov, V.V., Rupasov, A.A., Silin, V.P., Sklizkov, G.V., Tikhonchuk, V.T., Shikanov, A.S.; Zh. Eksp. Teor. Fiz. 67, 118(1974).

/3/ Ripin, B.H., McMahon, J.M., McLean, E.A., Manheimer, W.M., Stamper, J.A.; Phys. Rev. Lett. 33, 634(1974).

/4/ Yamanaka, C., Yamanaka, T., Sasaki, T., Mizui, J., Kang, H.B.; Phys. Rev. Lett. 32, 1038(1974).

/5/ Sigel, R., Eidman, K., Pant, H.C., Sachsenmaier, P.; Phys. Rev. Lett. 36, 1369(1976).

/6/ Massey, R., Berggren, K., Pietryzk, A.Z.; Phys. Rev. Lett. 36, 963(1976).

/7/ Offenberger, A.A., Cervenan, M.R., Yam, A.M., Pasternak, A.W.; J. Appl. Phys. 47, 1452(1976).

/8/ Donaldson, T.F., Hablard, M., Spalding, I.J.; Phys. Rev. Lett. 37, 1348(1976).

/9/ Grek, B., Pepin, H., Johnston, T.W., Leboeuf, J.N., Baldis, H.A.; Nucl. Fusion 17, 1165(1977).

/10/ Ripin, B.H., Young, E.C., Stamper, J.A., Armstrong, C.M., Decoste, R., McLean, E.A., Bodner, S.E.; Phys. Rev. Lett. 39, 611(1977).

/11/ Phillion, D.W., Kruer, W.L., Rupert, V.C.; Phys. Rev. Lett. 39, 1529(1977).

/12/ Burnett, N.H., Baldis, H.A., Enright, G.D., Richardson, M.C., Corkum, P.B.; J. Appl. Phys. 48, 3727(1977).

/13/ Ng, A., Pitt, L., Salzman, D., Offenberger, A.A.; Phys. Rev. Lett. 42, 307(1979).

/14/ Eidman, K., Brederlow, G., Brodman, R., Petsch, R., Sigel, R., Tsakiris, G., Volk, R., Witkowski, S.; J. Phys. D 12, 145(1979).

/15/ Ripin, B.H., McLean, E.A.; Appl. Phys. Lett. 34, 809(1979).

/16/ Herbst, M.J., Clayton, C.E., Chen, F.F.; Phys. Rev. Lett. 43, 1591(1979).

/17/ Gray, D.R., Murdoch, J., Sim, S.M.L., Cole, A.J., Evans, R.G., Toner, W.T.; Plasma Phys. 22, 967(1980).

/18/ Mayer, F.J., Busch, G.E., Kinzer, C.M., Estabrook, K.G.; Phys. Rev. Lett. 44, 1498(1980).

/19/ Herbst, M.J., Clayton, C.E., Peebles, W.A., Chen, F.F.; Phys. Fluids 23, 1319(1980).

/20/ Turner, R.E., Goldman, L.M.; Phys. Rev. Lett. 44, 400(1980). Phys. Fluids 24, 184(1981).

/21/ Artsimovich, V.L., Gorbonov, L.M., Kasyanov, Yu.S., Korobkin, V.V.; Zh. Eksp. Teor. Fiz 80, 1859(1981).

/22/ Turuchek, J.J., Chen, F.F.; Phys. Fluids 24, 1126(1981).

/23/ Handke, J., Rizvi, S.A.H., Kronast, B.; Appl. Phys. 25, 109(1981).

/24/ Walsh, C.J., Meyer, J., Hilko, B.; Appl. Phys. Lett. 38, 82(1981).

/25/ Ng, A., Offenberger, A.A., Karttunen, S.J.; Optics Comm. 36, 200(1981).

/26/ Tarvin, J.A., Schröder, R.J.; Phys. Rev. Lett. 47, 341(1981).

/27/ Decoste, R., Lavigne, P., Pepin, H., Mitchel, G.R., Kieffer, J.-C.; J. Appl. Phys. 53, 3505(1982).

/28/ Walsh, C.J., Meyer, J., Hilko, B., Bernard, J.E., Popil, R.; J. Appl. Phys. 53, 1409(1982).

/29/ Walsh, C.J., Baldis, H.A.; Phys. Rev. Lett. 48, 1483(1982).

/30/ Mitchel, G.R., Grek, B., Johnston, T.W., Pepin, H., Church, P., Lavigne, P., Martin, F., Decoste, R.; Phys. Rev. Lett. 48, 1472(1982).

/31/ Handke, J., Rizvi, S.A.H., Kronast, B.; Phys. Rev. Lett. 51, 1660(1983).

/32/ Baldis, H.A., Walsh, C.J.; Phys. Fluids 26, 3426(1983).

/33/ Clayton, Ch.E., Joshi, Ch., Chen, F.F.; Phys. Rev. Lett. 51, 1656(1983).

/34/ Tanaka, K., Goldman, L.M., Seka, W., Short, R.W., Williams, E.A.; Phys. Fluids 27, 2960(1984).

/35/ Herbst, M.J., Grun, J., Gardner, J., Stamper, J.A., Young, F.C., Obenschain, S.P., McLean, E.A., Ripin, B.H.; Phys. Rev. Lett. 52, 192(1984).

/36/ Aleksandrov, V.V., Brenner, M.V., Kovalsky, N.G., Loburyev, S.V., Rubenchik, A.M.; Zh. Eksp.Teor. Fiz. 88, 781(1985).

/37/ Huey, H.E., Mase, A., Luhmann, N.C., DiVergilio, W.F., Thomson, J.J.; Phys. Rev. Lett. 45, 795(1980).

/38/ Akiyama, H., Matsumoto, O., Takeda, S.; Proc. Int. Conf. Plasma Phys. Nagoya, Japan, 1980. V 1, 7 P-1-24.

/39/ Gorbunov, L.M.; Zh. Eksp. Teor. Fiz. 55, 2298(1968).

/40/ Drake, J., Kaw, P., Lee, Y.C., Schmidt, G., Liu, C.S., Rosenbluth, M.N.; Phys. Fluids 17, 778(1974).

/41/ Forslund, D.W., Kindel, J.M., Lindman, E.L.; Phys. Fluids 18, 1002(1975).

/42/ Bornatici, M.; J. Plasma Phys. 14, 105(1975).

/43/ Cohen, B.I., Max, C.E.; Phys. Fluids 22, 1115(1979).

/44/ Tang, C.C.; J. Appl. Phys. 37, 2954(1966).

/45/ Tang, C.L.; J. Appl. Phys. 40, 2510(1969).

/46/ Pesme, D., Laval, G., Pellat, R.; Phys. Rev. Lett. 31, 203(1973).

/47/ Harvey, R., Schmidt, G.; Phys. Fluids 18, 1395(1975).

/48/ Fuchs, V.; Phys. Fluids 19, 1554(1976).

/49/ Fuchs, V., Beaudry, G.; J. Math. Phys. 17, 208(1976).

/50/ Kruer, W.L.; Phys. Fluids 23, 1273(1980).

/51/ Speziale, T., McGrath, J.F., Berger, R.L.; Phys. Fluids 23, 1275(1980).

/52/ Speziale, T.; Phys. Fluids 25, 389(1982).

/53/ Galeev, A.A., Laval, G., O'Neil, T.M., Rosenbluth, M.N., Zagdeev, R.Z.; Pisma Zh. Eksp. Teor. Fiz. 17, 48(1973).

/54/ Speziale, T.; Phys. Fluids 27, 2583(1984).

/55/ Rosenbluth, M.N.; Phys. Rev. Lett. 29, 565(1972).

/56/ Liu, C.S., Rosenbluth, M.N., White, R.B.; Phys. Fluids 17, 1211(1974).

/57/ Silin, V.P., Starodub, A.N.; Pisma Zh. Eksp. Teor. Fiz. 23, 609(1976).

/58/ Manheimer, W.M., Colombant, D.G.; Phys. Fluids 24, 2319(1981).

/59/ Zozulya, A.A., Silin, V.P., Tikhonchuk, V.T.; Pisma Zh. Eksp. Teor. Fiz. 36, 111(1982).

/60/ Gorbunov, L.M., Polyanichev, A.N.; Zh. Eksp. Teor. Fiz. 74, 552(1978).

/61/ Gorbunov, L.M., Polyanichev, A.N.; Fizika Plazmy 5, 566(1979).

/62/ Kurin, V.V., Permitin, G.V.; Fizika Plazmy 5, 1084(1979).

/63/ Andreev, A.A., Shazev, A.N.; Fizika Plazmy 6, 1036(1980).

/64/ Gorbunov, L.M.; Zh. Eksp. Teor. Fiz. 81, 146(1982).

/65/ Dewandre, T., Albritton, J.R., Williams, E.A.; Phys. Fluids 24, 528(1981).

/66/ Johnston, T.W., Mitchel, G.R., Grek, B.; Phys. Fluids 25, 179(1982).

/67/ Mitchel, G.R., Johnston, T.W., Grek, B.; Phys. Fluids 25, 186(1982).

/68/ Berger, R.L.; Phys. Fluids 27, 1769(1984).

/69/ Randall, C.J., Thomson, J.J., Estabrook, K.G.; Phys. Rev. Lett. 43, 924(1979).

/70/ Randall, C.J., Albritton, J.A., Thomson, J.J.; Phys. Fluids 24, 1474(1981).
/71/ Short, R.W., Williams, E.A.; Phys. Rev. Lett. 47, 337(1981).
/72/ Short, R.W., Williams, E.A.; Phys. Fluids 26, 2342(1983).
/73/ Randall, C.J., Albritton, J.R.; Phys. Rev. Lett. 52, 1887(1984).
/74/ Baumgärtel, K., Sauer, K.; Phys. Rev. A26, 3031(1982).
/75/ Sauer, K., Baumgärtel, K.; Phys. Rev. Lett. 52, 1001(1984).
/76/ Mitchel, G.R., Johnston, T.W., Pepin, H.; Phys. Fluids 27, 1527(1984).
/77/ Evans, R.G.; Plasma Phys. 23, 99(1981).
/78/ Cairns, R.A.; Plasma Phys. 24, 109(1982).
/79/ Manheimer, W.M., Klein, H.H.; Phys. Fluids 17, 1889(1974).
/80/ Karttunen, S.J., Heikkinen, J.A.; Plasma Phys. 23, 869(1981).
/81/ Gorbunov, L.M.; Zh. Eksp. Teor. Fiz. 65, 990(1973).
/82/ Karabutov, A.A., Lapschin, E.A., Rudenko, O.V.; Zh. Eksp. Teor. Fiz. 71, 3(1976).
/83/ Karttunen, S.J., Salomaa, R.R.E.; Phys. Lett. A72, 336(1979).
/84/ Silin, V.P., Tikhonchuk, V.T.; Pisma Zh. Eksp. Teor. Fiz. 34, 385(1981).
/85/ Gorbunov, L.M., Polyanichev, A.N., Salikhov, D.K.; Fizika Plazmy 8, (1982).
/86/ Silin, V.P., Tikhonchuk, V.T.; Zh. Eksp. Teor. Fiz. 83, 1332(1982).
/87/ Gellert, B.; Opt. Comm. 46, 106(1983).
/88/ Heikkinen, J.A., Karttunen, S.J., Salomaa, R.R.E.; Phys. Lett. 101A, 217(1984). Phys. Fluids 27, 707(1984).
/89/ Karttunen, S.J., Salomaa, R.R.E.; Phys. Lett. 72A, 336(1979).
/90/ Karttunen, S.J.; Plasma Phys. 22, 151(1980).
/91/ Karttunen, S.J., McMullin, J.N., Offenberger, A.A.; Phys. Fluids 24, 447(1981).
/92/ Kruer, W.L., Valeo, E.J., Estabrook, K.G.; Phys. Rev. Lett. 35, 1076(1975).
/93/ Anderson, D., Wilhelmson; Nucl. Fusion 15, 387(1975).
/94/ Kruer, W.L., Estabrook, K.G., Sinz, K.H.; Nucl. Fusion 13, 952(1973).
/95/ Estabrook, K., Harte, J., Campbell, E., Ze, F., Phillion, D., Rosen, M., Larsen, J.; Phys. Rev. Lett. 46, 724(1981).
/96/ Gardner, J., Bodner, S.; Phys. Rev. Lett. 47, 1137(1981).
/97/ Montes, C.; Phys. Rev. Lett. 50, 1129(1983). Phys. Rev. A31, 2366(1985).
/98/ Armstrong, B.H., Decoste, R., McLean, E.A., Bodner, S.E.; Phys. Rev. Lett. 39, 611(1977).
/99/ Lehmberg, R.H.; Phys. Rev. Lett. 41, 863(1978).
/100/ Lehmberg, R.H., Holder, K.A.; Phys. Rev. A22, 2156(1980).
/101/ Lehmberg, R.H., Tripati, V.K., Liu, C.S.; Phys. Fluids 24, 703(1981).
/102/ Feldman, B.J., Bigio, I.J., Fisher, R.A., Phipps, C.R., Watkins, D.E., Thomas, S.J.; Los Alamos Science Vol. 3 Nr. 3, 2(1982).
/103/ Mitchel, G.R., Grek, B., Johnston, T.W., Pepin, H., Church, P., Martin, F.; J. Appl. Phys. 53, 3497(1982).

/104/ Klein, H.H., Manheimer, W.M., Ott, E.; Phys. Rev. Lett. 31, 1187(1973).

/105/ Manheimer, W.M., Klein, H.H.; Phys. Fluids 17 , 1889(1974).

/106/ Estabrook, K.G., Valeo, E.J., Kruer, W.L.; Phys. Lett. 49A , 109(1974).

/107/ Forslund, D.W., Kindel, J.M., Lindman, E.L.; Phys. Rev. Lett. 30 , 739(1973).

/108/ Forslund, D.W., Kindel, J.M., Lindman, E.L.; Phys. Fluids 18 , 1017(1975).

/109/ Forslund, D.W., Kindel, J.M., Lee, K., Lindman, E.L., Morse, R.; Phys. Rev. A11 , 679(1975).

/110/ Estabrook, K., Harte, J., Campbell, E.M., Ze, F., Phillion, D.W.; Phys. Rev. Lett. 46 , 724(1981).

/111/ McCall, S.L.; Appl. Phys. Lett. 32 , 284(1978).

/112/ Ikeda, K., Daido, H., Akimoto, O.; Phys. Rev. Lett. 45 , 709(1980).

/113/ Winful, H.G., Cooperman, G.D.; Appl. Phys. Lett. 40 , 298(1982).

/114/ Zozulja, A.A., Silin, V.P., Tikhonchuk, V.T.; Zh. Eksp. Teor. Fiz. 86 , 1296(1984).

/115/ Zozulja, A.A., Silin, V.P., Tikhonchuk, V.T.; Phys. Lett. 112A , 220(1985).

/116/ Zozulja, A.A., Silin, V.P., Tikhonchuk, V.T.; Fizika Plazmy 11 , 1344(1985).

/117/ Zaskalju, O.P., Zozulja, A.A., Kizylazow, Ju.I., Pamaisti, N.N., Silin, V.P., Tikhonchuk, V.T., Fabelinski, I.L.; Zh. Eksp. Teor. Fiz. 87 , 1581(1984).

10 Strongly nonlinear stationary waves

In this chapter we shall start to investigate wave-plasma interaction processes where the relative density changes produced by the ponderomotive force of the high-frequency field cannot longer be considered as small. Thus, attempts to describe such processes have to start from the fully nonlinear system (3.14) - (3.16). Since there exist no general analytical methods of solving such systems of coupled partial differential equations, progress is usually made by looking for certain special solutions. One method which is often used and gives valuable insight, is that of looking for nonlinear stationary waves. This may be done by introducing the plane wave ansatz, i. e., by introducing a variable $y = x - ut$ and assuming that all dependent variables depend on x and t only through the combination y :

$$\vec{E} = \vec{E}(y) , \quad N = N(y) , \quad \vec{v} = \vec{v}(y) . \tag{10.1}$$

Mathematically, this converts partial differential equations into ordinary differential equations. Although the ansatz (10.1) restricts the manifold of solutions substantially, there are some interesting solutions which give insight into the mechanisms that couple the plasma motion to the field structure and vice versa.

Equilibrium configurations of plasmas under the action of the radiation pressure of a s-polarized high-frequency field and the fluid pressure have been studied first by Volkov /1/. His investigations were extended to include plasma flow by Vaskov and Karpman /2/. In this paper, some types of nonlinear stationary waves were discussed qualitatively. In the review paper of Gorbunov /3/, where also stationary waves are studied, the connection between a standing envelope soliton and the nonlinear skin effect is pointed out. The latter effect arises when an electromagnetic wave of not too high intensity enters an overdense plasma half-space. Moreover, there is a close relationship between nonlinear stationary waves and the problem of self-consistent density profile structures at the incidence of a strong electromagnetic wave upon a nonuniform plasma. This was pointed out by several authors who studied density profile deformation by radiation pressure in the stationary limit /4 - 7/. Strongly nonlinear envelope solitons, particularly their restricted regions of existence with respect to the propagation velocity, were studied in /8 - 11/. Further, stationary nonlinear waves in form of relativistically coupled electromagnetic and Langmuir waves in a cold plasma were investigated in /12, 13/.

10.1 Basic equations

We start from the one-dimensional version of the system (3.14) - (3.16) and assume an essentially monochromatic transverse (s-polarized) electric field which coexists with a plasma whose density varies along the x-axis ($\vec{v}$ in x-direction, $\vec{E}_\omega$ perpendicular to the x-direction):

$$\frac{\partial N}{\partial t} + \frac{\partial (Nv)}{\partial x} = 0 , \tag{10.2}$$

$$\frac{\partial v}{\partial t} + v\frac{\partial v}{\partial x} = - c_s^2 \frac{\partial}{\partial x} \ln N - \frac{1}{4} c_s^2 \frac{\partial}{\partial x} \left| E_\omega \right|^2 , \tag{10.3}$$

$$2i\omega \frac{\partial E_\omega}{\partial t} + c^2 \frac{\partial^2 E_\omega}{\partial x^2} + \omega^2 \mathcal{E}(\omega, N) E_\omega = 0 . \tag{10.4}$$

In the wave Eq. (10.4) $E_\omega(x, t)$ represents the complex envelope of the electric field with respect to the variation in time after separating the basis frequency ω according to Eq. (3.10). $\mathcal{E}(\omega, N)$ denotes the dielectric constant given by Eq. (3.11). In the above equations E_ω is scaled by $\omega m_e v_T/e$ as introduced in Section 9.1.

It should be pointed out that we have to be cautious when applying the considerations of this chapter to a longitudinal high-frequency carrier field because of the limited validity of the longitudinal wave equation

$$2i\omega \frac{\partial E_\omega}{\partial t} + 3v_T^2 \frac{\partial^2 E_\omega}{\partial x^2} + \omega^2 \mathcal{E}(\omega, N) E_\omega = 0$$

in describing waves in regions where the plasma density is significantly below the critical density N_c. These regions are characterized by strong Landau damping and, unlike the electromagnetic case, there exist no longitudinal free space waves.

We now seek special solutions of the system (10.2) - (10.4) in terms of the independent variable

$$z = k_o y = k_o (x - ut) , \quad k_o = \omega / c , \tag{10.5}$$

that is, we want to find nonlinear waves stationary in a frame moving with velocity u.

Introducing the dimensionless real quantities $A(x, t)$, $\varphi(x, t)$ by the relation

$$E = A e^{i\varphi} , \tag{10.6}$$

the continuity Eq. (10.2) and the equation of motion (10.3) are converted by the ansatz (10.5) into the relations

$$N(v - u) = C_1 , \tag{10.7}$$

$$\frac{(v - u)^2}{2} + c_s^2 \ln N + \frac{1}{4} c_s^2 A^2 = C_2 , \tag{10.8}$$

C_1, C_2 being free constants. They can be fixed, if we impose the condition, that the density takes the value N_o in a flow-free and field-free region. This condition is to a certain extent artificially because one cannot know in advance whether a solution attains somewhere the values $v = 0$, $A = 0$ at all. From Eqs. (10.7), (10.8) it follows with $v = 0$, $A = 0$

$$- N_o u = C_1 , \quad \frac{1}{2} u^2 + c_s^2 \ln N_o = C_2 .$$

Introducing the dimensionless density

$$Q = N/N_o \tag{10.9}$$

and the Mach number

$$M = u/c_s , \tag{10.10}$$

Eqs. (10.7), (10.8) can be written in the form

$$Q\left(\frac{v}{c_s} - M \right) = - M , \tag{10.11}$$

$$\frac{1}{2} M^2 (1/Q^2 - 1) + \ln Q + \frac{1}{4} A^2 = 0 . \tag{10.12}$$

In the limit of small density deviations ($Q = 1 + q$, $|q| \ll 1$), Eq. (10.12) takes the form

$$(1 - M^2) q + \frac{1}{4} A^2 = 0 .$$

Introducing the plane wave ansatz (10.5) into the Eq. (10.4) for the envelope field yields the two ordinary differential equations

$$- 2 \frac{u}{c} A \frac{dA}{dz} + 2A \frac{dA}{dz} \frac{d\varphi}{dz} + A^2 \frac{d^2\varphi}{dz^2} = 0 , \tag{10.13}$$

$$\frac{d^2 A}{dz^2} + 2 \frac{u}{c} A \frac{d\varphi}{dz} - A \left(\frac{d\varphi}{dz} \right)^2 + (1 - Q/Q_c) A = 0 \tag{10.14}$$

with $Q_c = N_c/N_o$. After multiplying by A, Eq. (10.13) can be integrated once to give

$$\frac{d}{dz} \left\{ - \frac{u}{c} A^2 + A^2 \frac{d\varphi}{dz} \right\} = 0 ,$$

that is,

$$\frac{d\varphi}{dz} = \frac{u}{c} + \frac{\varphi_o}{A^2}$$

with another free constant φ_o. Setting $\varphi_o = 0$ at the moment, we get

$$\frac{d\varphi}{dz} = \frac{u}{c} . \tag{10.15}$$

Introducing this expression into Eq. (10.14) yields

$$\frac{d^2 A}{dz^2} + \left\{ 1 - \left(\frac{u}{c} \right)^2 - \frac{Q}{Q_c} \right\} A = 0 .$$

The term $(u/c)^2$ is small compared with one since u has the order of the ion acoustic velocity and $c_s/c \ll 1$. Strongly nonlinear stationary waves are then governed by the set of the following two equations

$$\frac{d^2A}{dz^2} + \left(1 - \frac{Q}{Q_c}\right)A = 0, \tag{10.16}$$

$$\frac{1}{2}M^2(1/Q^2 - 1) + \ln Q + \frac{1}{4}A^2 = 0. \tag{10.17}$$

The associated flow profile follows from Eq. (10.11). Free parameters entering the theory are the Mach number $M = u/c_s$ of the stationary wave and the ratio $Q_c = N_c/N_o$ of the critical density associated with the carrier wave frequency and the undisturbed density N_o (so far existing at all).

Allowing φ_o to be different from zero creates the additional term φ_o/A^3 in the wave Eq. (10.16). To illustrate the physical meaning of this constant, let us consider the simple case of two counterpropagating monochromatic waves with different amplitudes in a homogeneous medium. The corresponding field envelope E(x) can be written as

$$E(x) = e^{ikx} + f\,e^{-ikx + i\alpha}, \quad f \text{ real}.$$

Then the quantity $A = (E E^*)^{1/2}$ can easily be shown to obey the equation

$$\frac{d^2A}{dx^2} + k^2A - \frac{k^2(1-f^2)^2}{A^3} = 0. \tag{10.18}$$

This suggests that φ_o is a measure of the relative intensity of forward and backward propagating waves. With other words, φ_o is related to the Poynting vector of the high-frequency field and $\varphi_o = 0$ is equivalent to zero energy flow. We shall restrict ourselves to discuss this case only.

From Eq. (10.17) it follows

$$\frac{1}{2}M^2(1/Q^2 - 1) + \ln Q \gtrless 0 \tag{10.19}$$

as a necessary condition for the existence of a stationary wave. Fig. 10.1 illustrates this con-

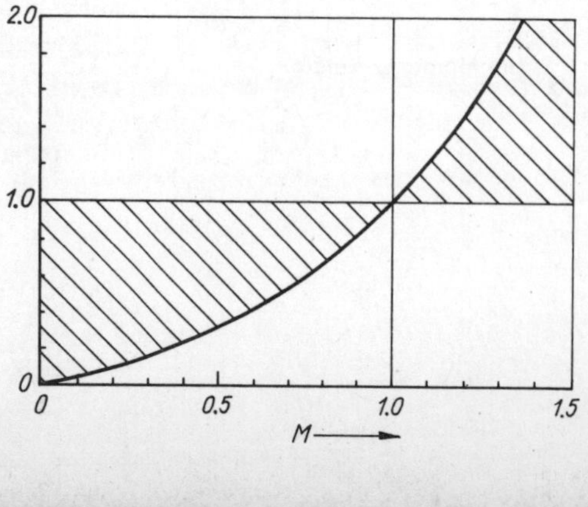

Fig. 10.1 Areas in the M-Q-plane (hatched) where stationary waves are allowed to exist according to Eq. (10.19)

dition, showing the areas (hatched) where Eq. (10.19) is fulfilled. From this graph it follows that a finite field amplitude is associated with a density depression (density hump) in a stationary wave, propagating with subsonic (supersonic) velocity.

In order to get an overview over the different types of stationary waves, we shall derive an equation for the phase trajectories $P = P(Q)$ $(P = dQ/dz)$ in the P-Q-plane and shall look how this diagram changes if the parameters M, Q_c are changed.

The differential equation governing the P-Q-phase diagram can easily be derived. From Eq. (10.17) it follows

$$A^2 = 2M^2(1 - 1/Q^2) - 4\ln Q\,. \tag{10.20}$$

After differentiating two times we get

$$Q(M^2 - Q^2)\frac{d^2Q}{dz^2} + \left(Q^2 - 3M^2 - 2\frac{(M^2 - Q^2)^2}{Q^2A^2}\right)\left(\frac{dQ}{dz}\right)^2 + \frac{Q^4A^2}{2}\left(1 - \frac{Q}{Q_c}\right) = 0\,,$$

and, using $P = dQ/dz$, we find

$$\frac{dP}{dQ} = -\frac{\left(Q^2 - 3M^2 \quad \frac{2(M^2 - Q^2)^2}{Q^2A^2}\right)P^2 + \frac{Q^4A^2}{2}\left(1 - \frac{Q}{Q_c}\right)}{PQ(M^2 - Q^2)}\,. \tag{10.21}$$

This equation can be supplemented by an explicit expression for $P = P(Q)$: After multiplying by dA/dz, Eq. (10.16) can be written as

$$\frac{d}{dz}\left[\frac{1}{2}\left(\frac{dA}{dz}\right)^2 + \frac{1}{2}A^2\right] - \frac{Q}{Q_c}\,\frac{d}{dz}\left(\frac{1}{2}A^2\right) = 0\,.$$

With the help of Eq. (10.17) the last term in this equation can also be written as a total derivative, giving

$$-Q\frac{dA^2}{dz} = 2\frac{d}{dz}(M^2/Q + Q)\,.$$

Therefore, we can write

$$\left(\frac{dA}{dz}\right)^2 + A^2 + 4(M^2/Q + Q)/Q_c = \text{const.} = C \tag{10.22}$$

and, using Eq. (10.20), we get finally

$$P^2 = \left(\frac{dQ}{dz}\right)^2 = \frac{A^2Q^6}{4(M^2 - Q^2)^2}\left[C - A^2 - \frac{4}{Q_c}(M^2/Q + Q)\right]\,. \tag{10.23}$$

In what follows, we draw conclusions on the qualitative properties of nonlinear stationary waves by studying the topology of the P-Q-phase diagram.

10.2 Equilibrium states (envelope solitons, nonlinear skin effect)

We shall look in this section for stationary waves that do not move relative to the plasma. With $u = 0$, i. e., $M = 0$, Eq. (10.17) simply gives

$$\frac{N}{N_o} = Q = \exp(-A^2/4) \tag{10.24}$$

and from the continuity Eq. (10.11) it follows $v = 0$. Obviously, there is no plasma flow associated with such equilibrium configurations. With the help of Eq. (10.24) the density Q can be eliminated from Eq. (10.16) and we obtain

$$\frac{d^2A}{dz^2} + \left[1 - \frac{1}{Q_c}\exp\left(-\frac{A^2}{4}\right)\right] A = 0\,. \tag{10.25}$$

This equation can be considered as a kind of generalized nonlinear Schrödinger equation, valid also for large field amplitudes A. Configurations described by Eq. (10.25) are characterized by a balance between the fluid pressure of the plasma and the radiation pressure of the high-frequency field. Setting $v = 0$, from the momentum Eq. (10.3) it follows

$$\frac{d}{dx}\left[\ln N + \frac{1}{4}A^2\right] = 0\,,\ \text{i.e.}\,,\ \ln N + \frac{1}{4}A^2 = \text{const.}$$

That is, the total pressure (sum of plasma pressure and radiation pressure) remains constant.

In Fig. 10.2, phase diagrams are shown for an overcritical (a) and a subcritical value (b) of the undisturbed density N_o. In the former case, there exist two types of periodic solutions, including the soliton as a limit case (described by the separatrix leaving the point $Q = 1$, $P = 0$ with a finite derivative dP/dQ). Common to all solutions with $N_o > N_c$ is the existence of regions where the plasma density exceeds the critical density. This is illustrated in Fig. 10.3, which shows spatial profiles of density and field for typical solutions in the case $N_o = 1,5\ N_c$. The density profiles look like mirror imagines of the field structure, expressing the pressure balance, mentioned above. The fixed point $Q = Q_c$ in the diagram (a) corresponds to a singular

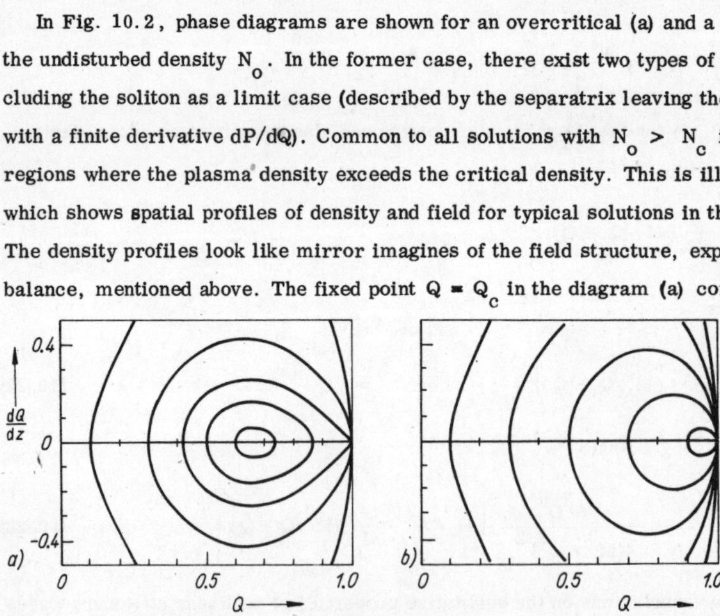

Fig. 10.2 Phase diagram $P = dQ/dz$ versus Q for stationary waves with $M = 0$; $N_o/N_c = 1.5$ (a), $N_o/N_c = 0.75$ (b)

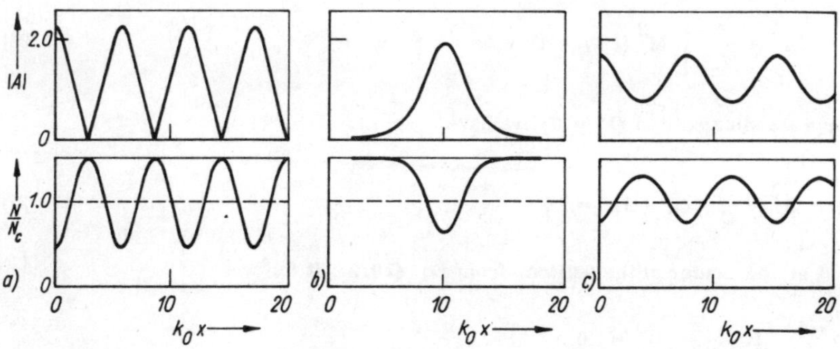

Fig. 10.3 Spatial profiles of field amplitude and density for stationary waves at rest (M = 0), corresponding to Fig. 10.2 (a)

solution without spatial variation of density and field in a plasma of exact critical density. Fig. 10.4 shows in addition spatial structures of solitons with increasing intensity. The relation between the extremum values A_{max}, Q_{min} of field amplitude and density at the center of the soliton and the parameter N_o/N_c can be expressed analytically. To this end we fix the constant C in Eq. (10.22) by imposing the boundary condition A, dA/dz going to zero for $Q \longrightarrow 1$, according to the property of a soliton solution. We find $C = 4(M^2 + 1)/Q_c$ and Eq. (10.22) then takes the form

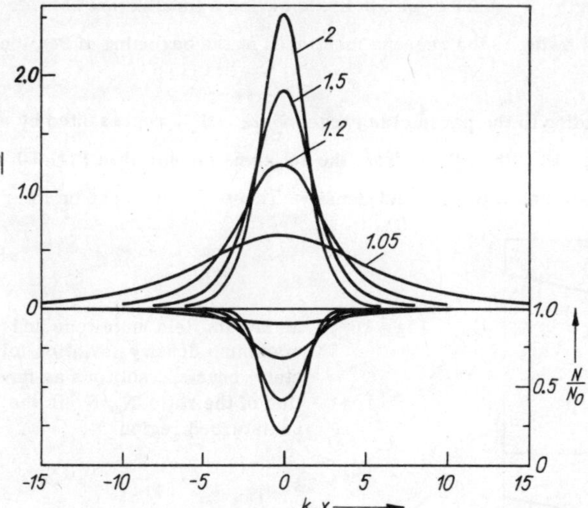

Fig. 10.4 Profiles of static envelope solitons of different amplitudes. Parameter is the ratio N_o/N_c of the undisturbed density to the critical density associated with the frequency of the carrier wave

$$\left(\frac{dA}{dz}\right)^2 + A^2 + \frac{4}{Q_c}\left[M^2(1/Q - 1) + Q - 1\right] = 0 . \tag{10.26}$$

In particular, for a standing soliton $(M = 0)$ we have

$$\left(\frac{dA}{dz}\right)^2 + A^2 + \frac{4}{Q_c}(Q - 1) = 0 . \tag{10.27}$$

Due to $dA/dz = 0$ at the center of the soliton, from Eq. (10.27) it follows

$$A^2_{max} + \frac{4}{Q_c}(Q_{min} - 1) = 0 .$$

Together with Eq. (10.24) this gives the desired relations

$$Q_c A^2_{max} + 4\left[\exp(-A^2_{max}/4) - 1\right] = 0 , \tag{10.28}$$

or

$$Q_c \ln Q_{min} + 1 - Q_{min} = 0 , \tag{10.29}$$

which are displayed in Fig. 10.5.

As may be seen from the figure, strong solitons require a fairly overcritical density in the undisturbed region. In the center of a strong soliton, the density is substantially reduced. Thus, with growing intensity solitons approach genuine cavities with a self-consistently trapped high-frequency field. The half-width of a strong static soliton with a carrier field of frequency ω has the order of the vacuum wavelength $\lambda_o = 2\pi c/\omega$. Solitons having a longitudinal carrier field are limited to small amplitudes owing to the reasons mentioned at the beginning of Section 10.1.

Static nonlinear waves corresponding to the parameter range $N_o < N_c$, represented by a phase diagram like that shown in Fig. 10.2 (b), differ from the solutions displayed in Fig. 10.3 in that the density remains everywhere below the critical density. These solutions can be inter-

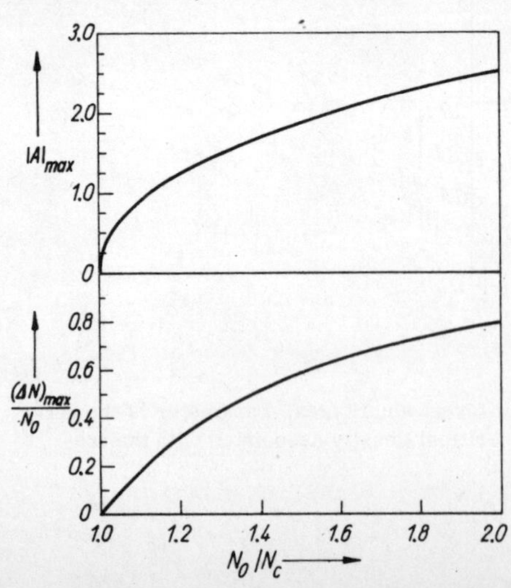

Fig. 10.5 Maximum field amplitude and maximum density deviation for static envelope solitons as function of the ratio N_o/N_c in the undisturbed region

preted in terms of a standing wave structure which coexists with a periodically modulated density profile. The periodicity interval is approximately given by half the wavelength of the carrier wave in a plasma of averaged density. Such configurations are similar to those found as final states when studying stimulated Mandelstam-Brillouin scattering as initial value problem (cf. Section 9.1).

Nonlinear skin effect

According to linear theory, an electromagnetic wave with $\omega < \omega_p$ is completely reflected from the plasma, penetrating it just to a distance of the order of the skin depth. Due to the spatial gradient in the field amplitude within the plasma, the ponderomotive force causes a density depression at the plasma boundary, allowing the field to penetrate deeper into the plasma. This effect, know as nonlinear skin effect, was pointed out by Silin /14/ (see also /3, 15 16/). Solitary equilibrium states as discussed above are closely related to this phenomenon.

The problem can be solved by fitting the vacuum field of the incident wave to a soliton profile cut at a certain position. Let an electromagnetic wave with frequency and normalized amplitude E_o be incident upon an overdense plasma half space. The corresponding vacuum field can be written as

$$E(x) = E_o(e^{iz} + Re^{-iz}), \qquad z \leq z_o, \; E_o \text{ real},$$

where R is the complex reflection coefficient and z_o is the position of the plasma boundary. We assume the plasma field to be represented by a soliton field, which, according to Eq. (10.6), may be written as

$$E(z) = A(z)\, e^{i\varphi} \quad , \qquad z \geq z_o .$$

The parameter $N_o/N_c = 1/Q_c$ which uniquely determines the soliton structure, is fixed by the density of the plasma half space and the frequency of the incident wave. Fitting both fields at the plasma boundary requires the continuity of E and dE/dz. This is expressed by the two equations

$$E_o(e^{iz_o} + Re^{-iz_o}) = A(z_o)\, e^{i\varphi} ,$$

$$iE_o(e^{iz_o} + Re^{-iz_o}) = \left.\frac{dA}{dz}\right/_{z=z_o} e^{i\varphi}$$

with the solution

$$R = \frac{dA/dz - iA}{dA/dz + iA}\, e^{2iz_o} , \tag{10.30}$$

$$E_o^2 = \frac{1}{4}\left\{ A^2 + \left(\frac{dA}{dz}\right)^2 \right\} . \tag{10.31}$$

Eq. (10.30) implies $|R| = 1$ due to A, dA/dz being real. The derivative dA/dz in Eq. (10.31) can be eliminated by Eq. (10.27) which results in the relation

$$E_o^2 = \frac{1 - Q(z_o)}{Q_c} = \frac{N_o}{N_c}(1 - Q(z_o)) . \tag{10.32}$$

This equation determines the position of the soliton profile, at which the fitting is possible. A necessary condition, however, for Eq. (10.32) to have a solution is

$$Q(z_o) \geq Q_{min}$$

where Q_{min} again denotes the minimum density of the soliton corresponding to the given value of N_o/N_c. Consequently, both solutions can be joined smoothly at z_o only, if E_o is within the range

$$0 < E_o^2 \leq \frac{N_o}{N_c}(1 - Q_{min}) . \tag{10.33}$$

This condition defines a threshold intensity $E_{o,\,thr}$ such that fitting fails for $E_o > E_{o,\,thr}$. Using the relations $Q_{min} = \exp(-A_{max}^2/4)$ and $E_{o,\,thr} = A_{max}/2$, the latter of which following from Eq. (10.31) with $dA/dz = 0$, we arrive at the implicit relation

$$E_{o,\,thr}^2 = \frac{N_o}{N_c}\left\{1 - \exp(-E_{o,\,thr}^2)\right\} \tag{10.34}$$

for this threshold intensity. For illustration of the dependence E_o as function of N_o/N_c we refer to Fig. 10.5 which shows $A_{max} = 2E_{o,\,thr}$ versus N_o/N_c.

In Fig. 10.6 spatial profiles of density and field amplitude are shown, illustrating the penetration of a wave in a fairly overcritical plasma ($N_o/N_c = 1.5$) for different intensities. In the limit of a small amplitude of the incident wave, the density is not affected and the amplitude decreases exponentially as predicted by the linear theory. The existence of the threshold value for E_o suggest a qualitatively different solution for incident intensities exceeding $E_{o,\,thr}$. We come back to this point in Chapter 11 where we shall see that in this case a steady-state solution does not longer exist.

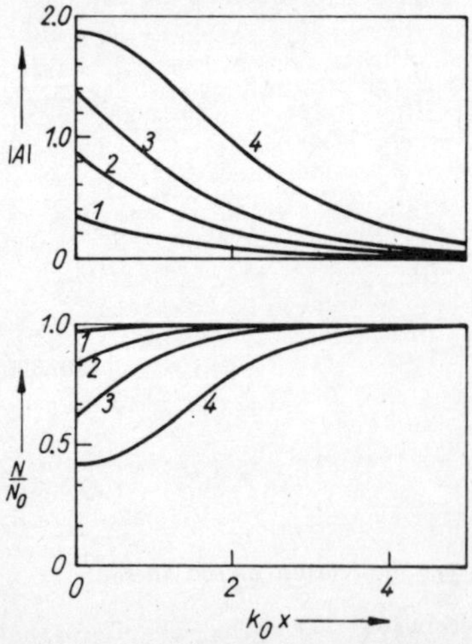

Fig. 10.6 Nonlinear skin effect: spatial profiles of field and density for increasing intensity of the incident wave; $E_o = 0.2$ (1), $E_o = 0.5$ (2), $E_o = 0.75$ (3), $E_o = 0.935$ (4); $N_o/N_c = 1.5$

10.3 Nonlinear stationary waves with plasma flow

Stationary waves involving plasma flow are described by solutions of the system (10.16), (1o.17) with $M \neq 0$. Fig. 10.7 shows typical phase plots in the case $N_o/N_c = 1.5$. The appearance of an additional separatrix (the line $Q = M$) gives rise to several complications. The most notable modification is the existence of a critical Mach number M_k that separates topologically different structures. Figs. 10.7 (a), (b), (c) represent the regions $M < M_k$, $M = M_k$ and $M > M_k$, respectively. A characteristic feature of the subsonic regime is that solitons continue to exist, again described by the separatrix emerging from the point $Q = 1$, $P = 0$ with finite derivative

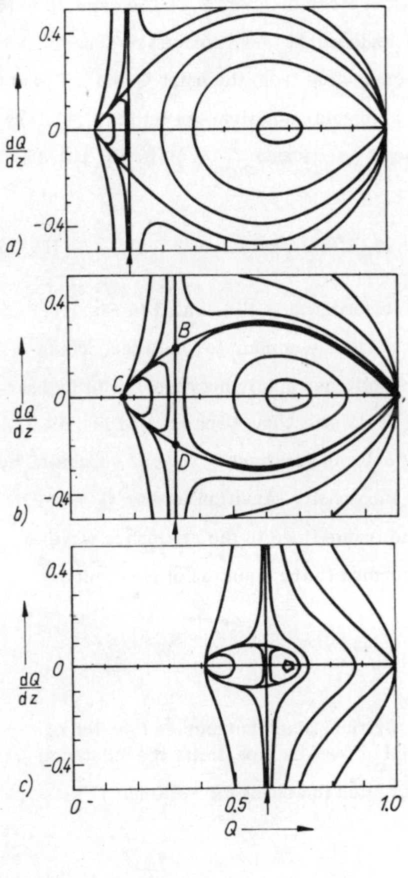

Fig. 10.7 Phase diagrams $P = dQ/dz$ versus Q for stationary waves with $M = 0.2$ (a), $M = M_k = 0.316$ (b) and $M = 0.6$ (c); $N_o/N_c = 1.5$. The arrow points to the separatrix $Q = M$

dP/dQ. The associated dependencies A_{max}, Q_{min} on N_o/N_c are modified by the finite Mach number M. An implicit relation can be derived from Eq. (10.26) with dA/dz = 0, using Eq. (10.17),

$$M^2 = \frac{2Q_{min}^2 \left[\ln Q_{min} + (1 - Q_{min}) Q_c \right]}{(Q_{min} - 1)(Q_{min} + 1 - 2Q_{min}/Q_c)} . \tag{10.35}$$

This formula can be found in /2/. From Fig. 10.7 (a) it is obvious that the minimum density of a soliton moving with Mach number M through the plasma satisfies the inequality $Q_{min} > M$. With growing M a configuration will be reached, where the separatrix corresponding to the soliton is identical to the separatrix crossing the line M = Q. In this situation we have Q_{min} = M and the associated Mach number represents the critical Mach number M_k. The case $M = M_k$ will be studied in more detail in the next section. If M exceeds M_k, solitons do no longer exist. This is seen from Fig. 10.7 (c): the separatrix emanating from the point Q = 1, P = 0 does not return to the axis P = 0 and thus represents a singular solution. An implicit relation for the critical Mach number M_k can simply be obtained by setting Q_{min} = M in Eq. (10.35). This results in the condition

$$(M_k - 1)(M_k + 1 - 2M_k/Q_c) - 2(\ln M_k + (1 - M_k)/Q_c) = 0 \tag{10.36}$$

which defines M_k as function of $N_o/N_c = 1/Q_c$. This dependence is illustrated in Fig. 10.8. The hatched region in this figure can be interpreted as existence domain for solitons. This diagram reveals that the maximum velocity allowed for solitons rapidly decreases with increasing N_o/N_c, that is, with increasing soliton amplitude. This effect has been studied in /8 - 11/.

Fig. 10.7 (a) indicates the existence of two types of subsonic stationary waves, separated by the line Q = M, thus differing in the range covered by the density variation (either Q < M or Q > M). The associated flow velocity $\tilde{v}$ relative to the frame fixed to the stationary wave is supersonic for Q < M and subsonic for Q > M, according to the equation of continuity

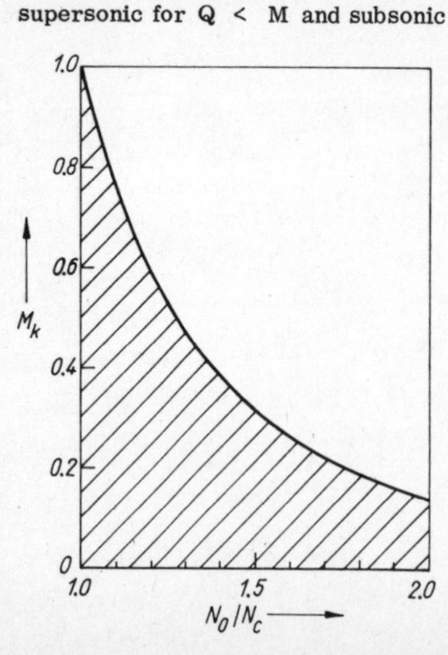

Fig. 10.8 Critical Mach number as function of N_o/N_c. The plot limits the existence region (hatched) for solitons

$$\frac{\tilde{v}}{c_s} = \frac{v}{c_s} - M = -\frac{M}{Q}.$$

The separatrix in Fig. 10.7 (a) which crosses the line $Q = M$ represents a particular solutio with alternating transitions between subsonic and supersonic flow.

10.4 Double-humped solitons, shock-like structures

The limit case $M = M_k$ is of particular interest, as we shall see below. The corresponding phase trajectory pattern is shown in Fig. 10.7 (b). The separatrix, which corresponds to the soliton for $M < M_k$ now crosses the line $Q = M$. Following it along the path A-B-C-D-A, the associated solution still represents a soliton-like solution because it exhibits a solitary density minimum. The amplitude profile, however, differs significantly from that of an ordinary soliton in having two maxima separated by a zero at the center. In Fig. 10.9, the amplitude structure and the profiles of density and flow velocity for such souble-humped soliton are shown.

The existence of this type of envelope solitons was first pointed out by Nishikawa et al. /17/ and Ikezi et al. /18, 19/. They observed the propagation of combined density-field structures of this kind with approximately ion acoustic velocity, carried by longitudinal plasma waves. Attempts to explain the phenomenon by the authors mentioned above were based on the Korteweg-de Vries equation, modified by including the ponderomotive force. Starting from similar equations, near-sonic double-humped solitions with a longitudinal basic wave have been treated numeri-

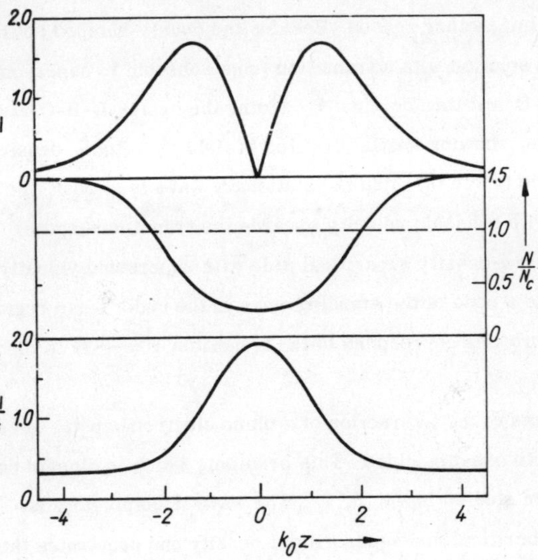

Fig. 10.9 Spatial profiles of field amplitude, density and plasma flow velocity for a double-humped soliton; $N_o/N_c = 1.5$

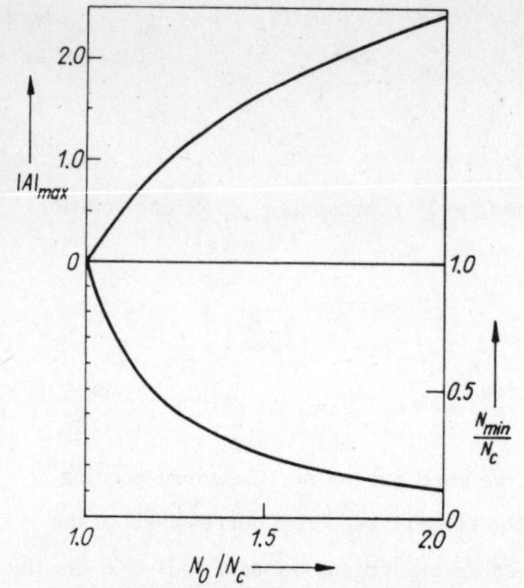

Fig. 10.10 Maximum field amplitude and density minimum of a double-humped soliton as function of N_o/N_c

cally by Bogolubsky and Makhankov /20, 21/, there denoted as ls-solitons.

Describing the phenomenon with the help of the KdV equation implies the presence of ion acoustic dispersion. It should be emphasized, however, that dispersion is not a crucial condition for double-humped solitons to exist as shown above. They prove to be a phenomenon solely created by the nonlinearity of the interaction between field and plasma.

Maximum amplitude and density minimum of a double-humped soliton are uniquely determined by the ratio N_o/N_c, the only free parameter. These dependendies are displayed in Fig. 10.10. Thus double-humped solitons represent a highly specialized subclass of the rich variety of solutions. Owing to the coexistence of a zero of the field amplitude with a density minimum, the double-humped soliton structure cannot longer be interpreted in terms of a balance between kinetic plasma pressure and radiation pressure.

The case $M = M_k$ is interesting also for another reason. Besides the double-humped soliton, the separatrix in Fig. 10.7 (c) provides a solution with a transition from subsonic to supersonic flow, if we first go along the path A-B-C-D and then continue by adding the cycles D-B-C-D ad infinitum. Fig. 10.11 illustrates this case, showing spatial profiles of field amplitude, density and flow velocity. Relative to the reference frame in which the stationary wave is at rest, the high-density overcritical plasma moves with subsonic velocity towards the transition region, passes the sonic point and expands on the low-density subcritical side with supersonic velocity. The maximum flow velocity corresponds to a node of the standing wave in the underdense region. The vertices in the density- and velocity profiles correspond to the transition D $\longrightarrow$ B in the phase diagram.

This solution can be interpreted in terms of the interaction of a plane electromagnetic wave with a freely expanding, overdense plasma in one dimension. This problem, which is closely related to laser-plasma interaction, has been studied in /4, 6, 7/. The wave is completely reflected at the transition region between subcritical and supercritical density and penetrates the plasma with subsonic velocity, leaving behind a supersonic plasma flow. Note, that again the whole flow pattern is determined by the choice of N_o/N_c, the only free parameter.

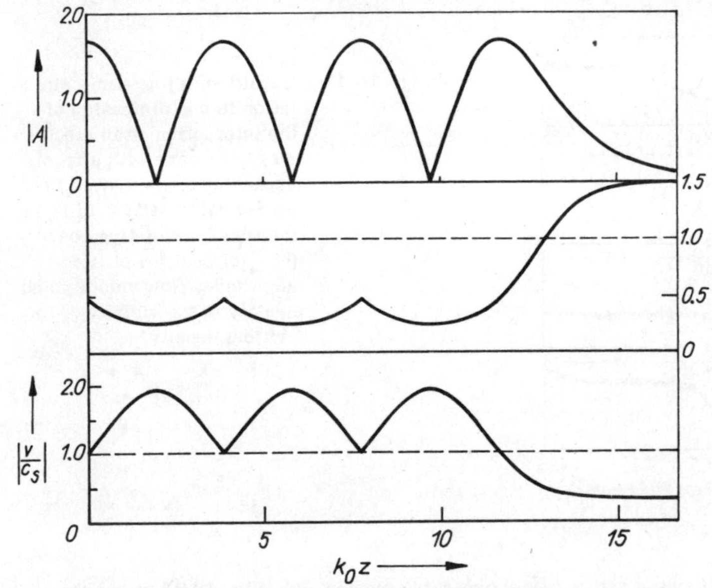

Fig. 10.11 Spatial profiles of field amplitude, density and plasma flow velocity for a shock-like structure, separating an overcritical field-free plasma with subsonic flow from an underdense supersonic plasma filled with a periodic field structure; $N_o/N_c = 1.5$

Fig. 10.12 illustrates the mutual relation between Mach number, upper shelf density, averaged values of flow velocity and lower shelf density with the maximum field amplitude A_{max} taken as independent variable. It is connected to the amplitude E_o of the incident wave approximately by the relation $A_{max} \simeq 2\,E_o/\sqrt{\varepsilon}$.

Shock-like profiles, as shown above have been found in a hybrid-code simulation of the inter-

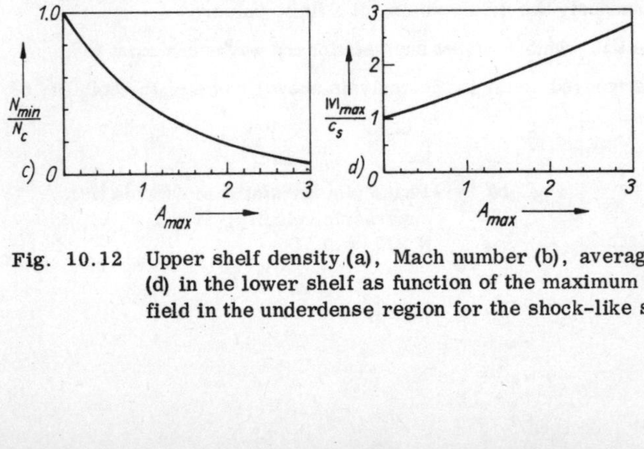

Fig. 10.12 Upper shelf density (a), Mach number (b), averaged density (c) and flow velocity (d) in the lower shelf as function of the maximum of the standing-wave electric field in the underdense region for the shock-like structure shown in Fig. 10.11

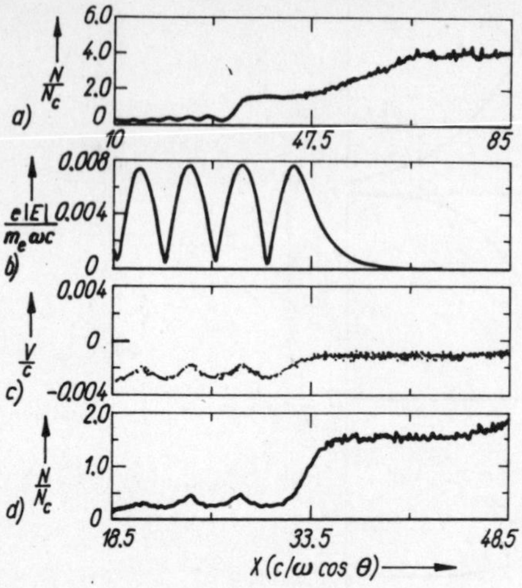

Fig. 10.13 Results of a long-time simulation in one dimension of the interaction of an electromagnetic wave with a freely expanding plasma after /4/. (a) Spatial density profile in the whole simulation box; (b) - (d) profiles of field amplitude, flow velocity and density in the vicinity of the critical density

action of an electromagnetic wave with a freely expanding plasma /4/. Fig. 10.13 shows the field structure and profiles of density and velocity as a result of the simulation. The electromagnetic wave is launched from the left boundary, finding an initial density profile which jumps from zero to four times the critical density at the middle of the simulation box. The formation of the upper density shelf and the modulated lower density shelf as a result of the interaction with the wave is clearly seen. Compared with free plasma expansion, this solution demonstrates the dramatic modification of the hydrodynamic behaviour of the plasma by the ponderomotive force of the electromagnetic field near the critical density.

Supersonic solitary waves

Looking for stationary waves propagating with supersonic velocity ($M > 1$), another solitary solution can be found. In the phase plot shown in Fig. 10.14 ($M = 1.5$, $N_o/N_c = 0.75$), this solitary solution is represented by the separatrix which connects the points $dQ/dx = 0$, $Q = 1$ and $dQ/dx = 0$, $Q = Q_c = N_c/N_o$. Associated spatial density profiles are displayed in Fig. 10.15. They are characterized by a zero of the field and a density minimum at the center. This solution, however, must be regarded as singular because the density approaches exactly the critical one at infinity and simultaneously the field remains at a finite value.

It should finally be emphasized that whether or not such stationary waves can exist in an actual plasma will depend on factors ignored so far in the analysis above; namely, the stability of

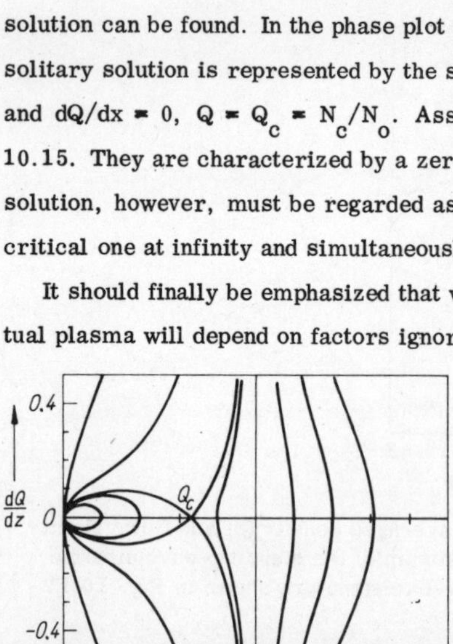

Fig. 10.14 Phase plot for stationary waves with supersonic velocity; $M = 1.5$, $N_o/N_c = 0.75$

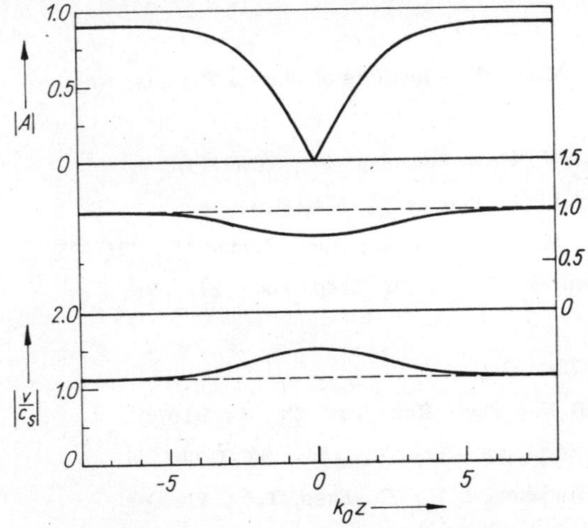

Fig. 10.15 Spatial profiles of field amplitude, density and flow velocity of a supersonic solitary wave; parameters as in Fig. 10.14

the solutions against various kinds of perturbations. An exception is the shock-like structure, the relevance of which to an initial value problem was pointed to. As far as solitions are concerned, stability has been proved by analytical as well as numerical methods within the frame of one space dimension /22 - 26/. There is, however, in general no stability against perturbations perpendicular to the propagation direction.

Some questions concerning the role of stationary waves as final states of a development in time will partly be answered in the next chapter which deals with instationary strongly nonlinear processes.

References

/1/ Volkov, T.V.; Plasma Physics (ed. Leontovich) Vol. 3, 395(1959).

/2/ Vaskov, V.V., Karpman, V.I.; Zh. Tekh. Fiz. 41, 676(1971).

/3/ Gorbunov, L.M.; Usp. Fiz. Nauk 109, 631(1973).

/4/ Lee, K., Forslund, D.W., Kindel, J.M., Lindman, E.L.; Phys. Fluids 20, 51(1977).

/5/ Sanmartin, J.R., Montanes, J.L.; J. Plasma Phys. 23, 349(1980).

/6/ Jones, R.D., Aldrich, C.H., Lee, K.; Phys. Fluids 24, 310(1981).

/7/ Aldrich, C.H., Jones, R.D., Lee, K.; Phys. Fluids 27, 2351(1984).

/8/ Shukla, P.K., Yu, M.Y., Tagare, S.G.; Z. Naturf. 31a, 1517(1976).

/9/ Schamel, H., Yu, M.Y., Shukla, P.K.; Phys. Fluids 20, 1286(1979).

/10/ Baumgärtel, K., Tskhakaya, D.D.; Beitr. Plasmaphysik 17, 75(1977).

/11/ Baumgärtel, K., Tskhakaya, D.D.; Phys. Lett. A62, 400(1977).

/12/ Akhiezer, A.I., Polovin, R.V.; Zh. Eksp. Teor. Fiz. 30, 915(1956).

/13/ Kaw, P.K., Sen, A., Valeo, E.J.; Physica 9D, 96(1983).

/14/ Silin, V.P.; Zh. Eksp. Teor. Fiz. 53, 1662(1967).

/15/ Sagdeev, R.Z.; in: Plasma Physics (ed. Leontovich) Vol. 3, 406(1959).

/16/ Sauer, K., Gorbunov, L.M.; Fizika Plazmy 3, 1302(1977).

/17/ Nishikawa, K., Hojo, H., Mima, K., Ikezi, H.; Phys. Rev. Lett. 33, 148(1974).

/18/ Ikezi, H., Nishikawa, K., Hojo, H., Mima, K.; Institute of Plasma Physics, Nagoya Univ. Report IPPJ - 169(1974).

/19/ Ikezi, H., Nishikawa, K., Mima, K.; J. Phys. Soc. Jap. 37, 766(1974).

/20/ Bogolubsky, I.L., Makhankov, V.G.; Fizika Plazmy 2, 974(1976).

/21/ Abdulloev, Kh.O., Bogolubsky, I.L., Makhankov, V.G.; Nucl. Fusion 15, 21(1975).

/22/ Gurevich, V.T., Karpman, V.I., Kaufman, R.N.; Zh. Eksp. Teor. Fiz. 56, 1979(1969).

/23/ Schmidt, G.; Phys. Rev. Lett. 34, 724(1975).

/24/ Denavit, J., Pereira, N.R., Sudan, R.N.; Phys. Rev. Lett. 33, 1435(1974).

/25/ Zakharov, V.E., Rubenchik, A.M.; Zh. Eksp. Teor. Fiz. 65, 997(1973).

/26/ Degtyarev, L.M., Zakharov, V.E., Rudakov, L.I.; Zh. Eksp. Teor. Fiz. 68, 115(1978).

11 Strongly nonlinear instationary interaction

The self-consistent development in space and time of an electromagnetic wave interacting with a uniform subcritical plasma has been studied at the lowest level of nonlinearity in Chapter 9 which deals with stimulated Mandelstam-Brillouin scattering. Under the assumption that the relative density perturbation induced by the ponderomotive force remains small, the interaction can be studied within the framework of a weakly nonlinear theory and the concept of mode coupling is appropriate. New features of the interaction process has been pointed out in the previous section, where a stationary model with a higher level of nonlinearity has been applied. The physical reason behind this is that the plasma reacts on a large amplitude field with the formation of strong spatial density nonuniformities, associated with plasma flow, which do no longer allow to describe the electromagnetic field in terms of a normal mode with slowly varying amplitude in space.

In this chapter we shall consider strongly nonlinear instationary processes associated with the interaction of an electromagnetic wave with a nonuniform plasma. This implies an extention of the theory of Chapter 9 to lift the approximation of weak nonlinearity and to include a zero-order plasma inhomogeneity as well as a generalization of the model of Chapter 10 to allow for time-dependence. We shall maintain, however, the restriction that all plasma quantities and the field vary only in one space dimension and that the plasma can be treated as isothermal fluid. We should further have in mind that in all of the formulas and figures the electric field will be normalized by $m_e \omega v_T/e$, i. e., $|E| = v_E/v_T$.

Our model is mathematically represented by the system (3.14) - (3.16) of coupled partial differential equations. We employ again the reference system introduced in Section 4.2 with the density gradient in x-direction and the x-z-plane as plane of incidence. In this orientation the electric field component E_y behaves independently from the other two components E_x, E_z. Eq. (3.16) splits into the three equations

$$2i\omega \frac{\partial E_x}{\partial t} + 3v_T^2 \frac{\partial^2 E_x}{\partial x^2} - ik_o c^2 \sin\theta \frac{\partial E_z}{\partial x} + \omega^2 (\mathcal{E} - \sin^2\theta) E_x = 0 , \tag{11.1}$$

$$2i\omega \frac{\partial E_y}{\partial t} + c^2 \frac{\partial^2 E_y}{\partial x^2} + \omega^2 (\mathcal{E} - \sin^2\theta) E_y = 0 , \tag{11.2}$$

$$2i\omega \frac{\partial E_z}{\partial t} + c^2 \frac{\partial^2 E_z}{\partial x^2} - ik_o c^2 \sin\theta \frac{\partial E_x}{\partial x} + \omega^2 \left(\mathcal{E} - \frac{3v_T^2}{c^2} \right) E_z = 0 , \tag{11.3}$$

the stationary version of which has already been used in Chapter 4 (Eqs. (4.22) - (4.24)). θ is the angle of incidence, k_o is the vacuum wave number of the incident wave and $\mathcal{E} = \mathcal{E}(N,\omega) = 1 - N/N_c (1 - i\nu/\omega)$ is the dielectric constant. The subscript ω, introduced to denote complex slowly time-dependent field quantities according to Eq. (3.10), has been dropped on all field components in the equations above. Eq. (11.2) describes the case of s-polarization ($E_x = E_z \equiv 0$) while the two coupled Eqs. (11.1), (11.3) apply for p-polarization ($E_y \equiv 0$). In either case we have to add the equations for the plasma dynamics

$$\frac{\partial N}{\partial t} + \frac{\partial}{\partial x}(N v) = 0 , \tag{11.4}$$

$$\frac{\partial v}{\partial t} + v \frac{\partial v}{\partial x} = - c_s^2 \frac{\partial}{\partial x} (\ln N + |\vec{E}|^2) . \tag{11.5}$$

The square of the absolute value of $\vec{E}$ in Eq. (11.5) should be read as

$$|\vec{E}|^2 = \begin{cases} |E_y|^2 & \text{for s-polarization} \\ |E_x|^2 + |E_z|^2 & \text{for p-polarization .} \end{cases}$$

In what follows we present results of a numerical study of Eqs. (11.1) - (11.5) for several initial configurations. Thereby problems are selected which are related to experimental observations on laser/microwave-plasma interaction or which allow to test the range of validity of the weakly nonlinear theory. Our focus is the modification of the density profile produced by the ponderomotive force associated with the high-frequency field in the plasma. This effect has been observed both in laser-target experiments /1 - 6/ and at microwave-plasma interaction /7 - 15/. In addition, the observations from recent ionospheric modification experiments /18, 19/ can be explained in terms of density profile deformation in form of caviton generation and trapping of Langmuir waves inside them.

The microwave experiments, carried out in large volume plasma devices, provide a rich variety of details on the density profile modification (see, for example /11/). Similar exhaustive information cannot be gained from laser-target experiments because these experiments cannot be diagnosed in detail.

It should first be noted that for density deformation of a nonuniform plasma to occur, the amplitude of the electric field is not required to be very large. As pointed out in Section 4.3, strong longitudinal fields are generated near the critical density by mode conversion in the case of p-polarization. The field may be enhanced by more than an order of magnitude and consequently ponderomotive forces may affect the density profile even for moderate levels of the strength of the incident wave.

The most significant effect arising when a p-polarized wave is incident on a nonuniform plasma is the generation of a density cavity and localized electric fields near the critical density in the early development. This effect is accompanied by the generation of fast electrons and ions /9/ (see also /16, 17/). The subsequent self-consistent evolution tends to form a density plateau with a steep rise from subcritical to supercritical density. This steepening affects resonance absorption by broadening the range of the angle of incidence for efficient absorption.

Several theoretical studies on density profile deformation at p-polarization are carried out within a steady-state model which reflects many of the important characteristics of the full time-dependent problem /20 - 23/. First calculations using a time-dependent model were based on the nonlinear Schrödinger equation including a capacitor field as an external source /24 - 28/. The model was subsequently improved by taking into account ion inertia. The self-consistent space-time development of density and field in the case when a wave with finite wave number is incident on a nonuniform plasma has been studied within the hydrodynamic model in /29 - 35/ and by means of particle simulation in /36 - 41/.

In the case of s-polarization, a higher level of the incident field amplitude is required to produce profile deformations. An orientation is given by the condition that the vacuum radiation pressure must be comparable to the kinetic plasma pressure. A hydrodynamic treatment of this interaction was given in /42 - 53/, partly in spherical geometry (see also references quoted in Chapter 10).

When a density cavity passes through a resonant structure in the sense of Chapter 5, electric field energy will temporarily be accumulated and subsequently reemitted, giving rise to an anomalous behaviour of the reflection coefficient (superreflection, /47/). This effect will be investigated in detail in Section 11.3. The fully nonlinear model will also be applied again to the problem of stimulated Mandelstam-Brillouin scattering in order to look for large-amplitude effects. In particular, the results of Section 9.7 on SBS saturation will be checked. Finally, the nonstationary evolution of an electromagnetic wave penetrating an overdense plasma half space and examples of plasma acceleration by an electromagnetic wave will be presented in Section 11.4.

The numerical procedure used to solve the equations for the plasma dynamics is the explicit SHASTA scheme developed by Boris and Book /54/. The boundary conditions imposed on density and velocity consist of either fixing the mass flux on both sides or fixing it at one side and allowing free-streaming out of the other end. The wave equation for the electric field has been discretizised according to the implicit Crank-Nicholson scheme /55/ with subsequent matrix inversion by the vector Gauss algorithm /56/.

For the calculations and presentation in figures a dimensionless time T has been introduced by

$$T = \omega_s t = 2k_o c_s t = 2 \frac{v_T}{c} \left(\frac{m_e}{m_i}\right)^{1/2} \omega t . \tag{11.6}$$

That is, the time is similarly scaled as in Chapter 9 by the ion acoustic frequency ω_s corresponding to a wave with $k_s = 2k_o$, k_o being the electromagnetic vacuum wave number.

11.1 Profile deformation at resonance absorption

The linear response of a plasma having a rigid nonuniform density profile on an incident p-polarized electromagnetic wave was discussed in Section 4.3. Near the critical density, the pump field is partly converted into short wavelength Langmuir waves propagating away from resonance point $\omega = \omega_p$ towards decreasing density. The level which is reached by the longitudinal field is determined by the group velocity of the Langmuir waves and will approximately satisfy /23/

$$\frac{|E_x|_{max}}{E_o} \simeq \frac{(k_o L)^{1/6}}{(v_T/c)^{2/3}}, \tag{11.7}$$

where L is the scale length of the density gradient and E_o, k_o refer to the amplitude and the wave number in absence of plasma. The electric field amplitude pattern illustrated in Fig. 4 may serve as a guide for the understanding of the nonlinear response of the system. It is predicted by Eqs. (11.1), (11.3) - (11.5) and involves the self-consistent distortion of the density profile under the ponderomotive action of the mode converted Langmuir waves. For conventional parameter combinations the longitudinal field maximum may exceed the vacuum amplitude E_o by more than a factor 10. That is, ponderomotive action are to be expected even for amplitudes $E_o \simeq 0.1$.

The problem has been treated in the past by several models having partly limited applicability. The stationary approximation has been used in /20 - 23/. A time-dependent model based on the nonlinear Schrödinger equation for the electric field parallel the density gradient was developed by Morales and Lee /24/ (see also /26, 27/). This treatment implies the assumption of an adiabatic connection between density deformation and field amplitude through the relation

$$\frac{\delta N}{N} = -\frac{1}{4} |E|^2,$$

i. e., the linearized version of Eq. (10.24). This model is capable of describing the formation of a density cavity and the associated spatial localization of the longitudinal field. It was completed by the same authors to include ion inertia and a detailed numerical analysis was given in the comprehensive paper /25/. This more general treatment confirmed the features of the earlier model and uncovered additional new effects as the excitation of oscillations of ion acoustic type after the formation of the density cavity and splitting of cavities into subcavities. In /33/ oscillations of the absorption coefficient have been found that are suggested to be produced by resonant configurations through which the density profile passes repeatedly. Self-consistent calculations in the approximation of a cold plasma have been performed in /29/. A study of profile deformation at p-polarization accompanied by simultaneous calculation of the second harmonic emission and the changes in the electron distribution function due to the absorbed wave energy has been given in /30 - 32/. As a characteristic feature, sharp peaks in the absorption have been found, associated with maximum harmonic emission. This effect is interpreted in terms of electrostatic structure resonance. It is further suggested that supersonic plasma flow near the critical density region tends to suppress the formation of density cavities.

In the following figures, results of a numerical treatment of Eqs. (11.1), (11.3) - (11.5) are displayed. Fig. 11.1 illustrates the early stage of the self-consistent space-time evolution of density and field when a p-polarized wave is incident on a plasma whose density varies linearly with distance at zero time. The main peak of the longitudinal field structure that exceeds the vacuum amplitude by a factor 6 in this case is responsible for the formation of the large cavity slightly below the critical density. As this cavity is formed, the excess density is expelled up and down the density gradient, forming a shoulder on the high-density side and giving rise to the generation of additional overcritical regions at the low density side. This process is accompanied by ion acoustic oscillations that are generated in response of the plasma to the density per-

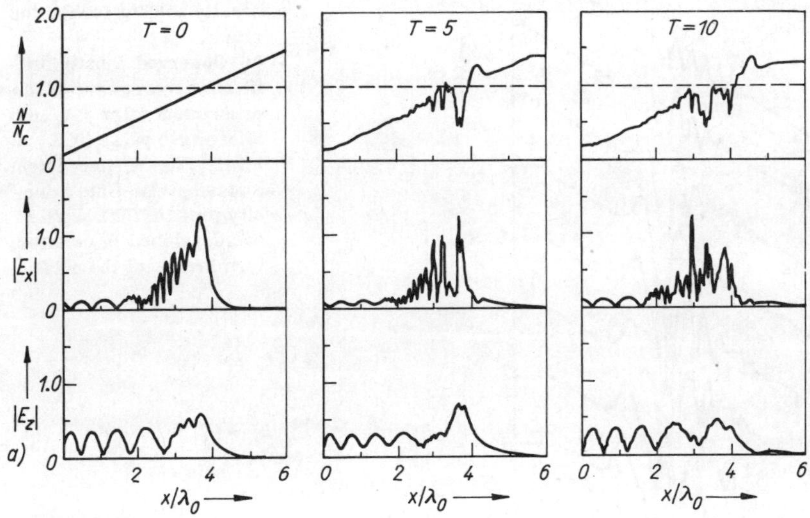

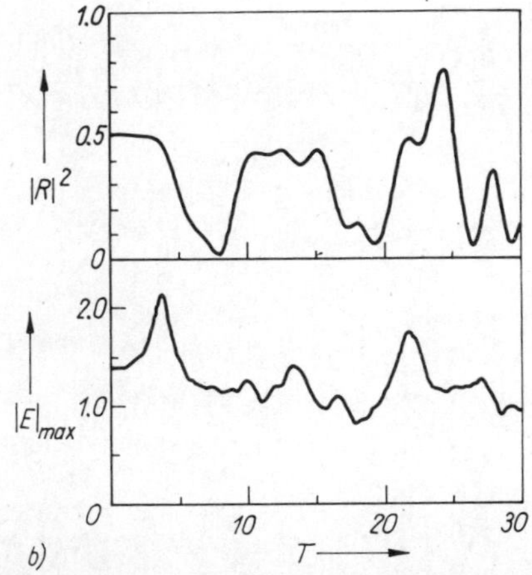

Fig. 11.1 Self-consistent deformation of a plasma having initially a linear density profile under the action of a p-polarized wave incident from the left.
(a) Spatial profiles of density and field components $|E_x|$, $|E_z|$ parallel and perpendicular to the density gradient at three times;
(b) reflection coefficient and maximum field amplitude versus time; $E_o = 0.2$, $L/\lambda_o = 4$, $\theta = 15^o$

turbation. To illustrate this point further, a sequence of density profiles separated by a smaller time step is shown in Fig. 11.2 (a). Another interesting effect is seen from this figure, namely the fission of the large cavity into two smaller subcavities. It seems that the lack of symmetry in the density profile with respect to the center of the cavity prevents the configuration from developing towards a steady state (a soliton). The longitudinal high-frequency energy begins to leak out on the low-density side of the cavity at a certain time leaving behind a smaller cavity.

Fig. 11.1 (b) complements the results, showing the time-evolution of the energy reflection coefficient and the maximum field amplitude within the space region covered by the calculation.

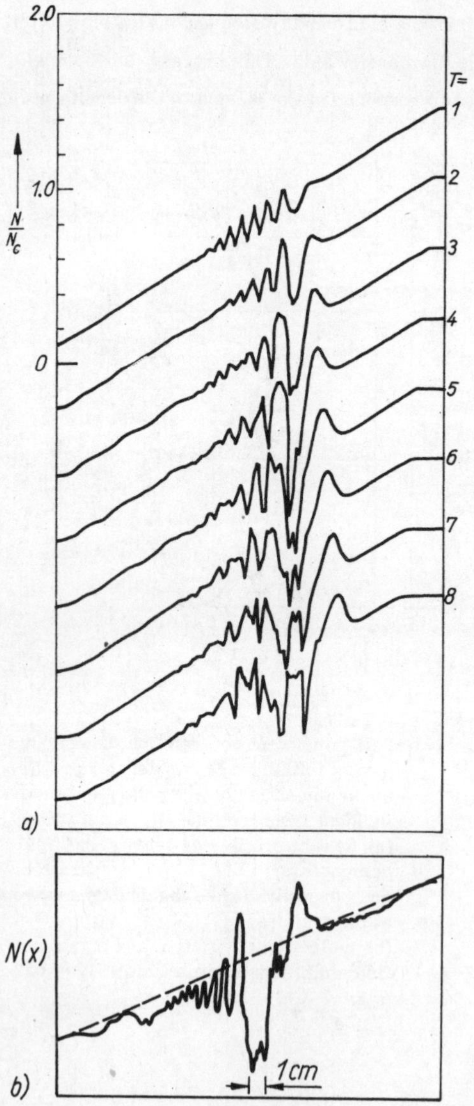

Fig. 11.2 (a) The same as Fig. 11.1; spatial density profiles illustrating the early stage of the profile deformation; density scaling for the top graph.
(b) Observed density profile in a microwave-plasma experiment after /11/. A microwave pulse (0.2 μs, 1 GHz, 10 kW) irradiates a plasma with linear density profile. The profile has developed 5 μs after termination of the pulse

The peaks in $|E|_{max}$ indicate that longitudinal structure resonances are temporarily attained. The significant drops in the reflection coefficient, however, are not associated with field maxima and can only be interpreted in terms of plateau-type resonances (cf. Fig. 11.3). The results displayed in Figs. 11.1, 11.2 indicate, generally speaking, a lost of coherence in the process with growing time. This reflects the complexity of the self-consistent interaction between plasma and field, consisting of a simultaneous operation of mode conversion, cavity formation and ion-acoustic-type response.

Most of the general features predicted by our model, represented by the Eqs. (11.1), (11.3)-(11.5), can also be extracted from the capacitor-field approximation of Morales and Lee /25/.

For comparision with experiment, Fig. 11.2 (b) displays the density structure observed at a plasma produced in a large volume device which was exposed to a microwave pulse (with an linear density profile at zero time). Essential features of this profile are reproduced by our model as shown in Fig. 11.2 (a).

When the plasma can be considered to be cold, the previous results are changed with some respect. Our model provides a view also on this altered situation. We arrive at the cold-plasma limit by setting $v_T/c = 0$ in the Eqs. (11.2), (11.3).

Linear theory predicts that we need to introduce a collisional damping rate in the case $v_T = 0$ if we want the longitudinal field to remain finite. The counterpart of Eq. (11.7) is then /23/

$$\frac{|E_x|\ max}{E_o} \simeq \frac{(k_oL)^{-1/2}}{\nu/\omega} \tag{11.8}$$

While, according to Eq. (11.7), the field enhancement is a slowly increasing function of the density gradient scale length in a plasma dominated by thermal effects, the opposite is true for a cold plasma as expressed by Eq. (11.8).

Density profile deformation in the case when thermal effects on the calculation of the electric field can be neglected is illustrated in Fig. 11.3. The self-consistent development proceeds in a modified way towards the formation of a subcritical plateau with a steep rise to the overcritical density regime. Thereby the plateau length increases with time. The formation of such a density plateau may give rise to the excitation of a certain type of structure resonance involving surface waves. This type was studied in Section 5.2 for the simple case of a two-level density profile. The field is amplified over an extended spatial region leading to enhanced absorption even for a low damping rate. This is illustrated in Fig. 11.3 (b) which exhibits the time evolution of the energy absorption coefficient. After a certain time, an absorption level fairly above the 50 %-limit predicted by resonance absorption is reached indicating the presence of a structure resonance.

For a qualitative comparison with experiment, observed density profiles at microwave-plasma interaction in the QUIPS device (Wong /11/) are displayed in Fig. 11.4 which show the plateau formation near the critical density.

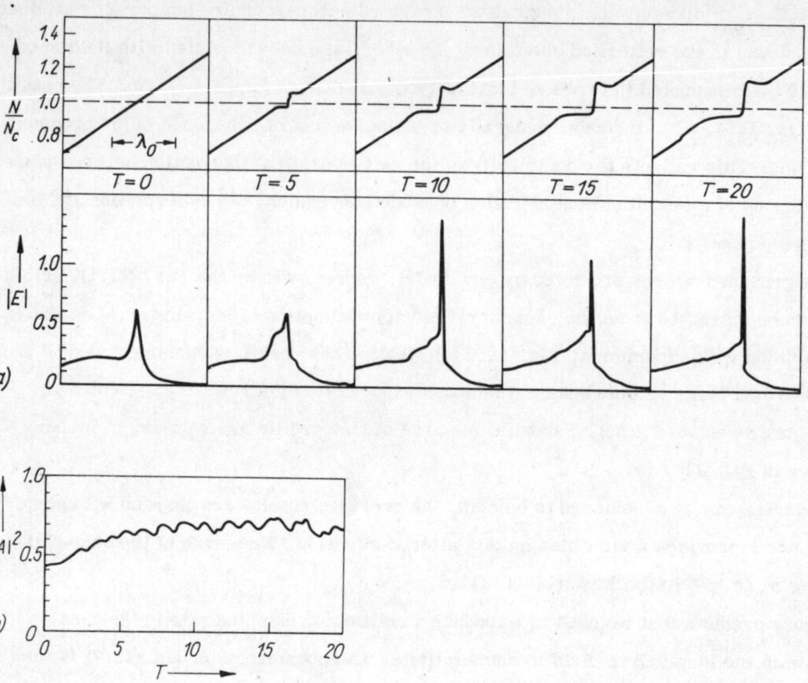

Fig. 11.3 The same as Fig. 11.1 for a cold plasma with finite electron-ion collision frequency. (a) Spatial profiles of density and the magnitude $|E| = (|E_x|^2 + |E_z|^2)^{1/2}$ of the electric field; (b) time history of the absorption. $E_o = 0.1$, $L = 4\lambda_o$, $\theta = 20^o$

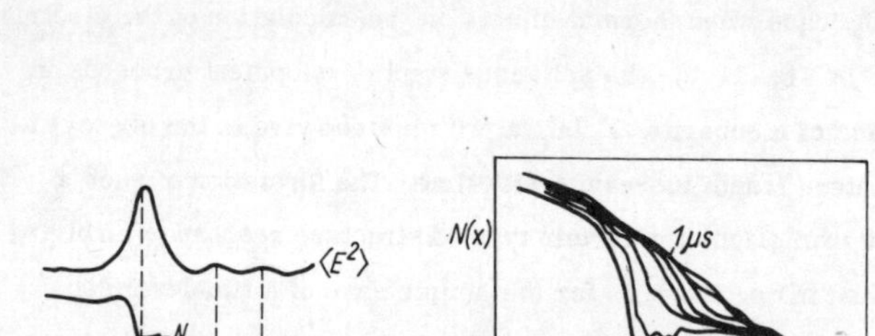

Fig. 11.4 Observed density plateau formation at microwave-plasma experiments after Wong /11/.
(a) Spatial profiles of field amplitude and density induced by a pulse of 10 W, 300 MHz in an argon-plasma 8 μs after termination;
(b) density profiles in a xenon-plasma produced by pulses of increasing length (1 μs - 8 μs)

11.2 Nonlinear stage of stimulated Mandelstam-Brillouin scattering

In this section we apply the fully nonlinear model to some problems pertaining to stimulated Mandelstam-Brillouin scattering which cannot be treated rigorously within the mode coupling approach. In particular, we shall examine zero-frequency effects on SBS, saturation of SBS by ion acoustic nonlinearities and the influence of plasma inhomogeneities combined with plasma flow on the space-time development of SBS. Especially the situation is of interest, in which the flow velocity matches the ion acoustic velocity somewhere along a nonuniform plasma flow profile.

Because the focus is here on subcritical density regions and effects occurring at the critical zone are excluded, the polarization of the incident wave is of less relevance and we restrict ourselves to p-polarization with propagation parallel to the density gradient. The field is then governed by the wave equation

$$2i\omega \frac{\partial E}{\partial t} + c^2 \frac{\partial^2 E}{\partial x^2} + \omega^2 \varepsilon E = 0 \tag{11.9}$$

which results from Eq. (11.2) by setting $\theta = 0$ and dropping the index y. It should be noted that the equation of motion (11.5) for the plasma is free of dissipative terms, thus linear ion acoustic damping is not included here.

As a preliminary result, Fig. 11.5 illustrates the space-time evolution of SBS in a uniform plasma of finite length ($L = 25\,\lambda_o$, $N = 0.3\,N_c$) when a strong electromagnetic wave ($E_o = 0.4$)

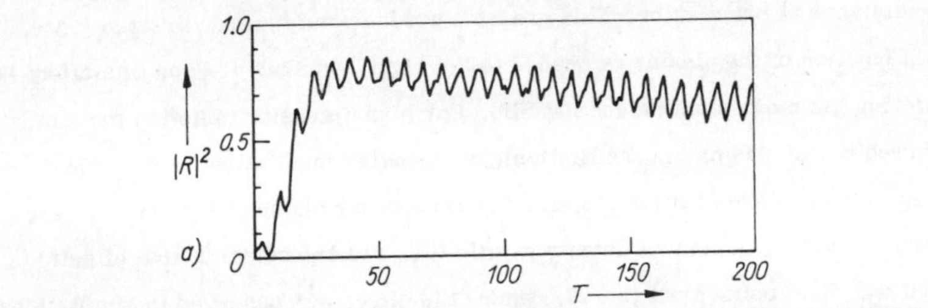

Fig. 11.5 Space-time development of stimulated Mandelstam-Brillouin scattering in a homogeneous plasma of finite extent.
(a) Reflectivity versus time, (b), (c) spatial profiles of density and field amplitude at two times. $E_o = 0.4$, $L/\lambda_o = 25$, $N_o/N_c = 0.3$

is incident from the left. A density jump to 0.5 N_c has been assumed at the right-hand edge of the plasma which causes a weak reflection at zero time sufficient to initiate the scattering process without the assumption of an initial ion acoustic noise level. The back reflection rapidly rises to a value of $|R|^2 \simeq 0.7$. The reflection coefficient, the time dependence of which is shown in Fig. 11.5 (a), suffers from weak oscillations with the ion acoustic period, resulting from the superposition of the directly reflected and the frequency-shifted SBS-reflected fractions of the pump wave. This effect was already found using the strong mode coupling scheme in Section 9.6 as a consequence of the presence of a partially reflecting boundary.

Note that the reflection coefficient does not exceed the 80 %-level. While $|R|^2$ is predicted to rise rapidly to the saturated value one by the mode coupling theory without ion acoustic damping for the same parameter combination (even for $E_o = 0.1$), the above result agrees reasonably with the prediction of the model which includes nonlinear ion acoustic damping (Section 9.7). From Fig. 9.24 we find, for example, a saturated level of $|R|^2 = 80\,\%$ for the parameters of Fig. 11.5 which combine to give a value $Y = 7.7$. Thus, ion acoustic nonlinearities are thought to be responsible for the SBS saturation exhibited in Fig. 11.5. This is supported by the spatial density profiles shown in Fig. 11.5 (b), (c). The wave fronts have no longer sinusoidal shape but suffer from steepening in direction of propagation. In addition, a phase change in the density perturbation with growing distance is indicated. The magnitude of the relative density disturbance is seen to be limited to $|\delta N/N| \lesssim 0.3$.

Another interesting effect can be seen from the density profiles displayed in Fig. 11.5. The spatially averaged density is suppressed at the side where the wave is incident. This can be understood as action of the ponderomotive force of the spatially averaged field structure which decreases with the distance of penetration. This gradient produces an inward-directed force, and consequently, a fraction of the plasma is pushed to the right. Another view on this effect is provided by considering the momentum balance at SBS. For high-intensity radiation the momentum that is transferred to the plasma via the Brillouin backscatter mechanism is not only carried away by the ion acoustic waves but affects also the zero-order plasma behaviour. In producing a non-uniform steady-state plasma density profile this effect provides a kind of self limitation of SBS. It was first pointed out in /57/ where the effect was observed in simulations.

Fig 11.6 refers to a situation where a supersonic plasma flow is present in a direction opposite to the pump wave direction. Such situations were studied in detail in Section 9.8 within the mode coupling theory. The time evolution of the scattering process was shown to be strongly affected by a feedback of the scattered wave and an oscillatory behaviour was found when the feedback factor defined by Eq. (9.94) attains a negative value. This case is realized by a proper choice of the parameters in Fig. 11.6. Pulsations of the backscattered intensity are indeed seen to dominate the evolution of SBS under such conditions. This can be regarded as confirmation of the results predicted by the mode coupling theory. The pulsation period is in accordance with Eq. (9.121) and the pulsations disappear if the parameters are changed such that the feedback factor is no longer real and negative.

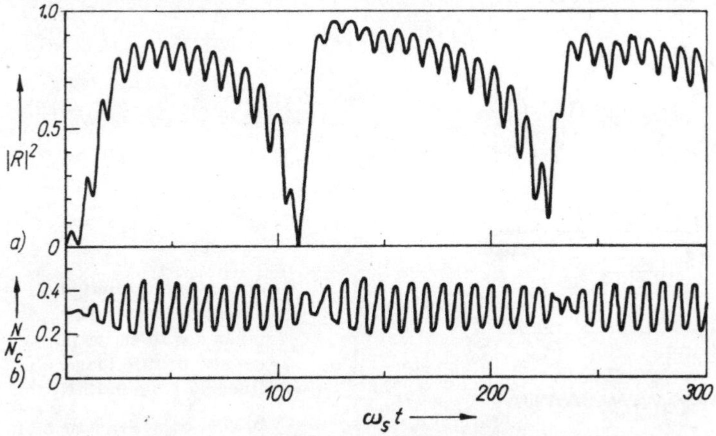

Fig. 11.6 Regenerative pulsations of the SBS reflectivity in the presence of a supersonic plasma flow against the direction of the pump wave as a result of feedback effects. Reflection coefficient (a) and density near the left-hand boundary (b) versus time. $E_o = 0.4$, $L/\lambda_o = 25$, $M = -2$, $N_o/N_c = 0.3$

Fig. 11.6 (b) displays the time evolution of the fluctuations in density at a fixed space point located near the left-hand side of the plasma. It is seen that the local density oscillation undergoes repeatedly a phase jump just at that time where $|R|^2$ goes through zero. This agrees with the model of the feedback mechanism developed in Section 9.5. This model is further supported by the spatial density profiles and the field structures desplayed in Fig. 11.7, corresponding to two different times where $|R|^2$ passes through a maximum (a) and through zero (b). The field structure shown in (b) reproduces the behaviour that should be expected according to Fig. 9.29 in reaching its maximum within the plasma.

In Figs. 11.8 and 11.9 the time evolution of SBS in a non-uniform plasma is illustrated. Fig. 11.8 represents a situation without plasma flow, showing that the development of the scattering process is only little affected by the plasma inhomogeneity in the early stage. At later times

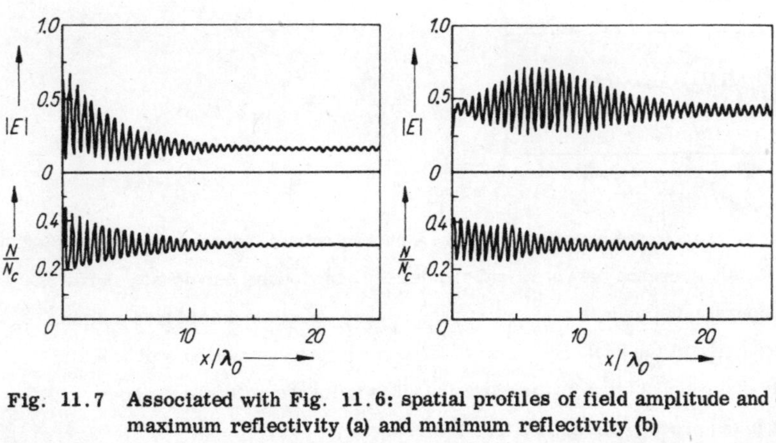

Fig. 11.7 Associated with Fig. 11.6: spatial profiles of field amplitude and density at maximum reflectivity (a) and minimum reflectivity (b)

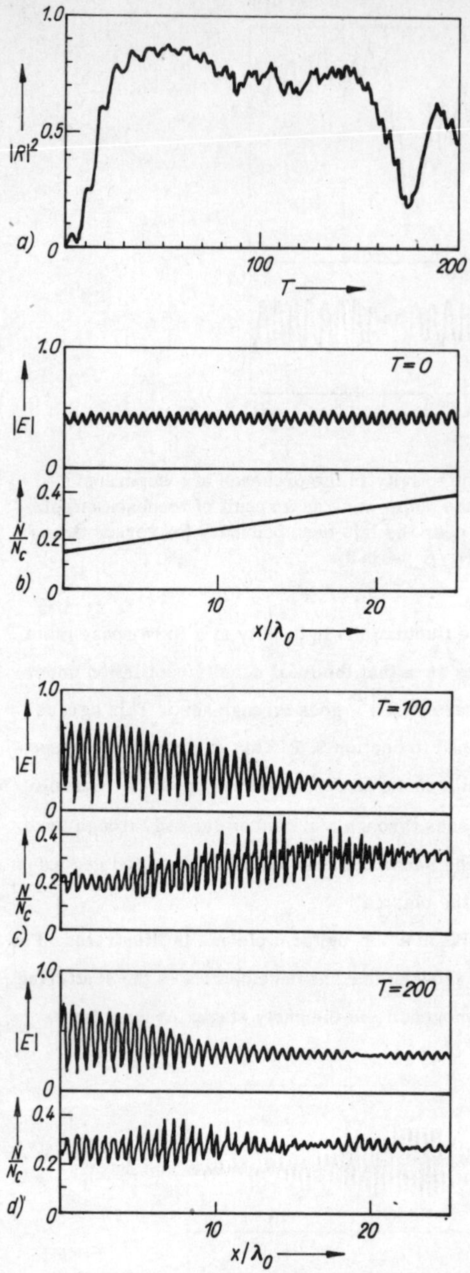

Fig. 11.8. Space-time evolution of SBS in a nonuniform plasma with an initial density profile increasing linearly from 0.15 N_c to 0.4 N_c over a distance of 25 vacuum wavelengths. (a) Reflectivity versus time; (b) - (d) spatial profiles of field amplitude and density for three times; $E_o = 0.4$

the reflectivity decreases (note the broad minimum). A characteristic feature common to Figs. 11.8 (c), (d) is a loss of coherence between field and density. Obviously, the density perturbation changes its phase with distance at a fixed time and tends to a rather disordered motion. This results in a reduction of the back-reflected signal by coherent distruction which in turn prevents the density disturbance from further growing. The influence of feedback, however, cannot be ruled out in the process shown in Fig. 11.8.

Finally, Fig. 11.9 illustrates the development of SBS in a nonuniform plasma in the pre-

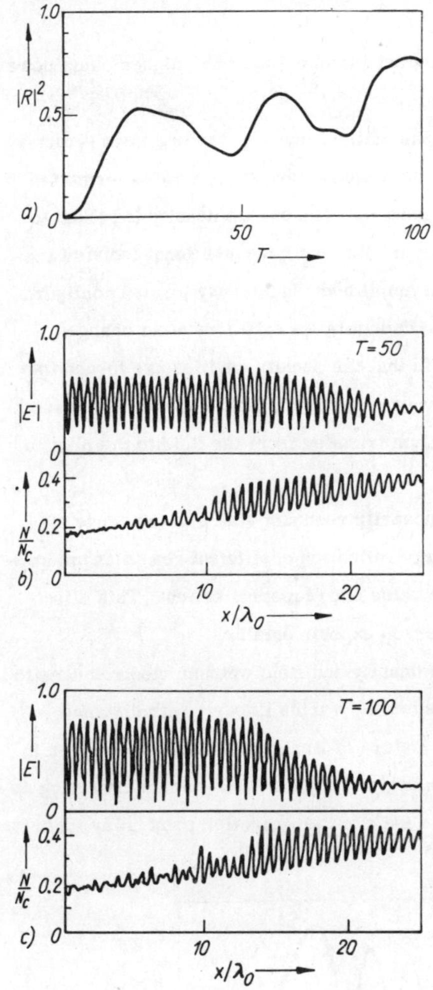

Fig. 11.9 The same as Fig. 11.8 in the presence of a plasma flow down the density gradient. A constant mass flux has been assumed at $T = 0$ such that the Mach number covers the interval $-2 \lesssim M \lesssim -0.75$ containing the sonic point $M = -1$

sence of plasma flow directed towards lower density. A constant mass flux along the plasma has been assumed at zero time such that the sonic point $v = c_s$ is within the interaction region. In contrast to the former case the reflection coefficient does not tend to decrease after the initial growth. The spatial density structures shown in Fig. 11.9 indicate that the operation regime of the SBS is concentrated around the location of the sonic point where the group velocity of the ion acoustic disturbances is approximately zero and an almost stationary Bragg grating can be established.

11.3 Superreflection

In this section we consider density profile deformation when a high-intensity wave ($E_o \gtrsim 1$) encounters a strongly nonuniform ($L/\lambda_o \lesssim 5$) plasma whose density rises to supercritical values. This problem can be considered as an extention of the investigations on stimulated

Brillouin scattering in a in homogeneous plasma to include strong gradients, higher pump power and the presence of a completely reflecting boundary.

Owing to the total reflection of the pump wave at the critical layer a standing wave structure is present even at zero time and the early development is characterized by a rapid formation of density cavities centered around the field amplitude maxima. The excess density is kicked to both sides of the cavities and the re-arranged density profile may have additional isolated spatial overcritical regions around the nodes of the field amplitude. This newly formed configuration, however, is far from being a steady-state one. It undergoes self-consistent changes which resemble the Brillouin scattering mechanism in that the density cavities are forced to move towards higher density with a velocity of the order of the ion acoustic speed. From another point of view, this is a consequence of the momentum transfer from the field to the plasma due to the total reflection.

With a certain probability the cavities attain temporarily resonant states in the sense of Section 5.1 during the time evolution. If so, the energy reflection coefficient reacts anomalously on such an event, exceeding for a certain time the value one (superreflection). This effect was first predicted in /47/ but has not yet been observed experimentally.

The early stage of the space-time development of density and field when an electromagnetic wave with $E_0 = 1.5$ is incident upon a plasma whose density varies linearly with distance $(L = 4\lambda_0)$ is illustrated in Fig. 11.10. The typical feature of Bragg reflection can be seen in this figure. The maxima of the standing wave pattern begin to dig cavities which start acting as Bragg grating for the incident wave and which cause a shift of the reflection point away from the

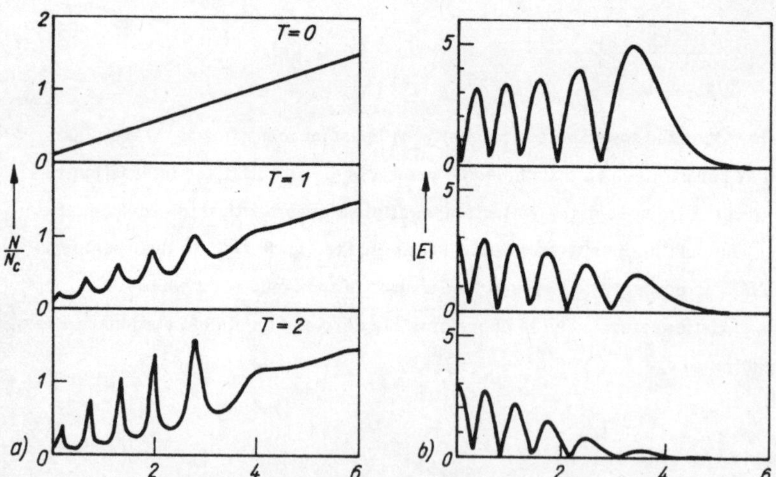

Fig. 11.10 Early stage of the self-consistent deformation of a linear plasma density profile under the action of a normal incident strong electromagnetic wave. Spatial profiles of density (a) and field amplitude (b) at three times. $E_0 = 1.5$, $L/\lambda_0 = 4$

critical layer to the lower density side. This is indicated by the decrease of the field maxima with penetration distance. A considerable amount of wave energy is therefore reflected in the subcritical region via the Bragg mechanism. Of course, these changes in the internal structure cannot be seen in the behaviour of the reflection coefficient because total reflection is maintained.

If looking at the time history of the reflection coefficient for longer times, however, we see rapid changes with significant deviations from the value one. A typical feature is that $|R|^2$ passes first through a minimum and afterwards jumps to a value greater than one (up to $|R|^2 \lesssim 3$). This means that temporarily the reflected energy is higher then the incident one. Fig. 11.11 displays this behaviour (parameters identical with that of Fig. 11.10). The time dependence of both the reflection coefficient (a) and the maximum amplitude (b) attained in the space region covered by the calculations is shown. Obviously, minima in $|R|^2$ coincide with peaks in $|E|$.

Fig. 11.11 (c) exhibits space profiles of density and field for a time at which $|E|$ passes through a maximum. One of the cavities is seen to be resonant in the sense of Section 5.1, the field trapped in it is appreciably enhanced. The anomalous variation of the reflection coefficient with time associated with a density cavity going through a structure resonance can be explained as follows: When the resonant state is approached, additional energy is required to build up the amplified electric field in the cavity. This energy is extracted from the incident wave, thus, the reflection coefficient drops below one. After the resonance is destroyed, the trapped high-frequency energy is released leading to a jump of $|R|^2$ to values greater than one, followed by

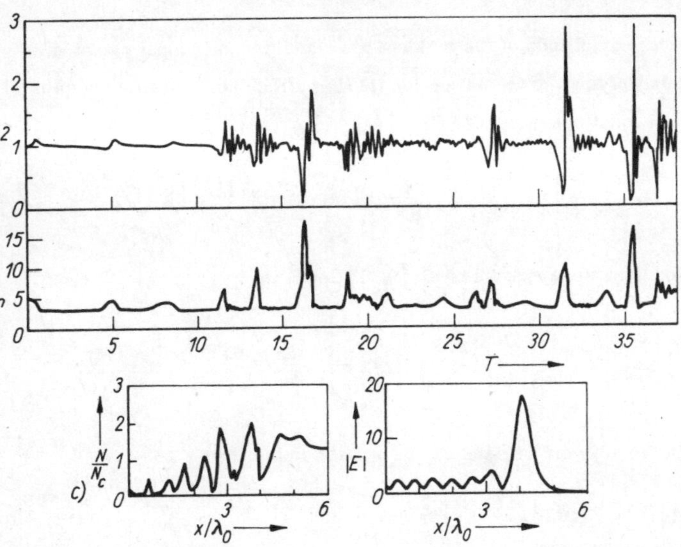

Fig. 11.11 Associated with Fig. 11.10; reflection coefficient (a) and field maximum (b) versus time; (c) spatial profiles of density and field at T = 16.3, at which an electromagnetic structure resonance is passed

relaxation oscillations. The occurrence of isolated peaks in the temporary evolution of the reflection coefficient is typical for plasmas with a strong initial density gradient. For weaker density gradients, the number of cavitons increases and so does the number of events described above, leading to rather a random variation of the reflection coefficient.

In the presence of collisional damping enhanced absorption is associated with resonantly amplified electromagnetic fields (up to 50 %). In this connection we point to an experiment /10/ in which a rapid drop in the reflection coefficient was observed when a overcritical plasma of finite extent was exposed to a large-amplitude microwave. The observed delay between switching-on the wave and the set in of the reduction in reflection agrees reasonably with the characteristic time for the formation of a resonant density cavity /52/.

Values $|R|^2 > 1$ are consistent with the wave equation (11.9). To illustrate this point we derive from this equation a conservation law which relates reflection, absorption and transmission.

Let $[0, d]$ be the spatial region covering the density profile. The coefficients of reflection, absorption and transmission are then defined as

$$|R|^2 = \left| \frac{E(0, t)}{E_0} - 1 \right|^2 ,$$

$$|T|^2 = \left| \frac{E(d, t)}{E_0} \right|^2 ,$$

$$|A|^2 = \frac{\nu}{\omega} \int_0^{k_0 d} \frac{N}{N} \frac{|E|^2}{E_0^2} \, d(k_0 x)$$

where E_0 and ν refer to the amplitude of the incident wave and the collisional damping decrement, respectively. Multiplying the wave equation (11.9) by E^*, the conjugate complex of Eq. (11.9) by E and adding both equations yields

$$2i\omega \frac{\partial}{\partial t} |E|^2 + c^2 \frac{\partial}{\partial x} \left[E^* \frac{\partial E}{\partial x} - E \frac{\partial E^*}{\partial x} \right] + 2i \frac{\nu}{\omega} \omega^2 \frac{N}{N_c} |E|^2 = 0 .$$

Integration of this equation over the space interval $[0, d]$ leads to

$$2i\omega \frac{\partial}{\partial t} \int_0^d |E|^2 dx + c^2 \left[E^* \frac{\partial E}{\partial x} - E \frac{\partial E^*}{\partial x} \right] + 2i \frac{\nu}{\omega} \omega^2 \int_0^d \frac{N}{N_c} |E|^2 dx = 0 . \tag{11.10}$$

Supposing that the field in the adjacent regions can be written in the form

$$E(x, t) = E_0 (e^{ik_0 x} + R(t) e^{-ik_0 x}) \quad \text{for} \quad x \leq 0 ,$$

$$E(x, t) = E_0 T(t) e^{ik_0 x} \quad \text{for} \quad x \geq d ,$$

the conditions for continuity of the field E and its spatial derivative dE/dx at the boundaries

$x = 0, d$ give the relation

$$\left[E^* \frac{\partial E}{\partial x} - E \frac{\partial E^*}{\partial x} \right]_0^d = 2ik_o E_o^2 \left\{ |T|^2 + |R|^2 - 1 \right\} . \tag{11.11}$$

Combining Eqs. (11.10) and (11.11) we arrive at the desired balance relation

$$\frac{\partial}{\partial(\omega t)} \int_0^{k_o d} \frac{|E|^2}{E_o^2} d(k_o x) + |R|^2 + |T|^2 + |A|^2 - 1 = 0 . \tag{11.12}$$

The conventional relation

$$|R|^2 + |T|^2 + |A|^2 = 1$$

is contained in Eq. (11.12) as stationary limit. $|R|^2$ being greater than one requires the first term in Eq. (11.12) to become negative. This is indeed the case during that time when the resonant field structure breaks down as a consequence of the cavity having lost its resonance property. The balance equation (11.12) has been checked numerically and found to be fulfilled.

Caviton model for superreflection

The principal mechanism underlying the superreflection can be isolated from the self-consistent development and can be studied in a rather simple model. In this model, the self-consistently formed density cavity is replaced by a stepwise uniform density profile consisting of two overcritical plasma layers separated by a subcritical region (see Fig. 5.7). The time development of the cavity is simulated by assuming that the width of the subcritical layer varies slowly with time. ($d_2 = d_{20} + ut$). The parameters are arranged such that the cavity passes a resonant state at a certain time.

The numerical solution of the weakly time-dependent wave equation (11.2) corresponding to the artificially constructed time-dependent density profile discussed above is illustrated in Fig. 11.12. The time history of both the reflection coefficient and the field amplitude maximum is shown. On passing the resonance the reflection coefficient shows a similar behaviour as in the

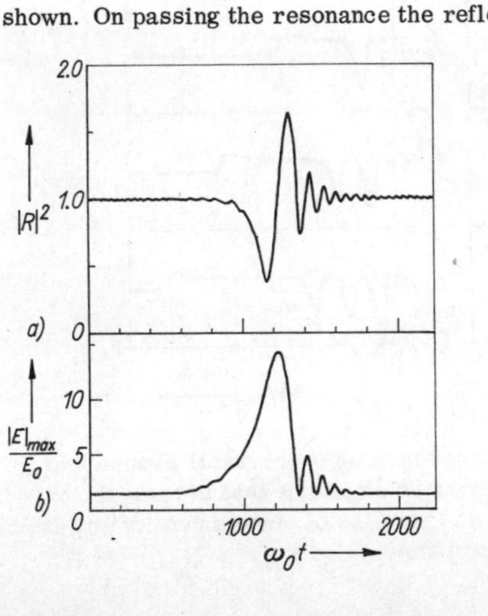

Fig. 11.12 Demonstration of superreflection within the simple caviton model, consisting of a stepwise uniform density profile as displayed in Fig. 5.7. The width of the cavity varies linearly with time according to $d_2 = 0.2\ \lambda_o + u\,t$ where $u/c = 5.5\ 10^{-4}$ and covers the interval $0.2 \leq d_2/\lambda_o = 0.3$. The cavity resonance is attained at $d_2 = d_2^{(1)} = 0.25\,\lambda_o$

self-consistent calculations. The significant deviation of $|R|^2$ from one coincides with the strong enhancemant of the field amplitude. Note the relaxation oscillations in $|R|^2$ after the main peak which are purely electromagnetic in nature.

The simple step-wise uniform density profile can also be treated analytically, even taking into account a finite damping rate /58, 59/. An interesting result is that the reflection coefficient cannot exceed a certain value about 3, no matter how the relevant parameters are chosen. This is in accordance with the numerical calculations.

11.4 Non-stationary skin effect; plasma acceleration; plasma confinement

In this last section we briefly consider another phenomena related to interaction of s-polarized waves with plasmas. The first one is concerned to the nonlinear skin effect. In Section 10.2 it was shown that the nonlinear skinning of an electromagnetic wave in an overcritical plasma half-space of density N can proceed in a stationary way only when the incident wave amplitude is below a certain threshold depending on the ratio N/N_c. Suppose now we have the opposite situation. This case has been treated by numerically solving Eqs. (11.4), (11.5), (11.9) for parameters which represent the region $E_o > E_{o,thr}$ ($E_o = 1.5$, $N/N_c = 1.5$). Fig. 11.13 illustrates the peculiarities of the interaction in this case. It is seen that the standing wave structure acts like a piston that compresses the plasma and initiates a shock wave moving ahead.

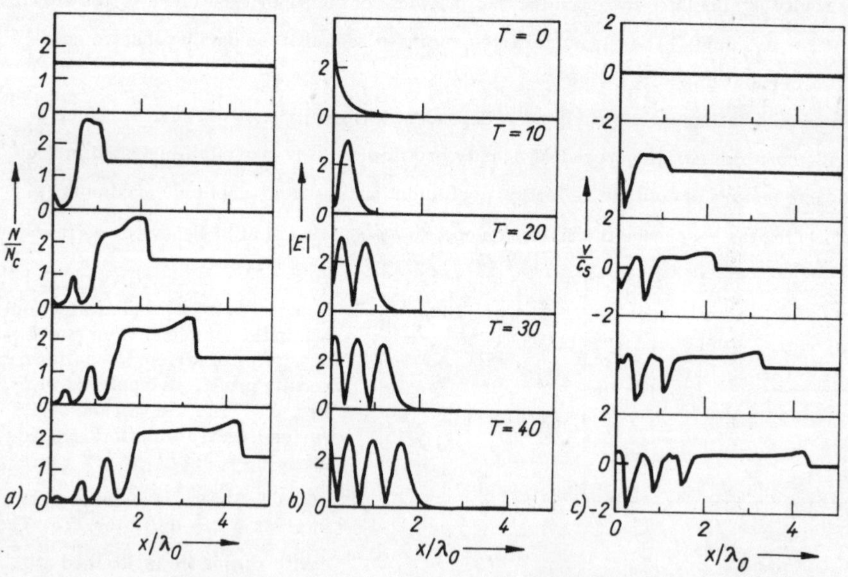

Fig. 11.13 Time-dependent nonlinear skin effect penetration of an overcritical plasma half space by a strong electromagnetic wave having an amplitude that exceeds the critical amplitude defined by Eq. (10.34). Spatial profiles of plasma density (a), field amplitude (b) and flow velocity (c) at different times; $E_o = 1.5$, $N_o/N_c = 1.5$

The velocity of the compressed plasma satisfies approximately the relation expressed by Fig. 10.12. A certain fraction of the plasma flows against the incident wave filling the regions of low radiation pressure. With growing time the configuration resembles the shock-like stationary wave exhibited in Fig. 10.11. Runs with lower amplitude of the incident wave predict solutions which merge into stationary configurations of the type shown in Fig. 10.6, thus confirming the existence of the threshold amplitude.

The second example demonstrates how a plasma of finite extent in x-direction will be deformed and accelerated by a large-amplitude wave. This problem has been treated analytically in /60/ under some limiting conditions for a plasma with a Gaussian density profile. Neglecting the kinetic pressure of the plasma and assuming the vacuum wavelength to be large compared with the half-width of the profile a uniform acceleration proportional E_o^2 results, accompanied by a compression of the profile.

Fig. 11.14 illustrates the result of a numerical calculation which simulates the case where an electromagnetic wave with amplitude $E_o = 2$ encounters a plasma with a Gaussian shape of the density profile (half-width $0.75\ \lambda_o$, maximum density $2\ N_c$). The profile is steepened and deformed and suffers from a constant acceleration (the center proceeds in space approximately proportional to t^2). The acceleration overcompensates the natural relaxation of the profile due to the density gradient and thus is responsible for keeping the profile overcritical. In case of lower amplitudes the maximum plasma density may fall below the critical density due to relaxation. Then the wave is transmitted by the profile and the acceleration is ceased.

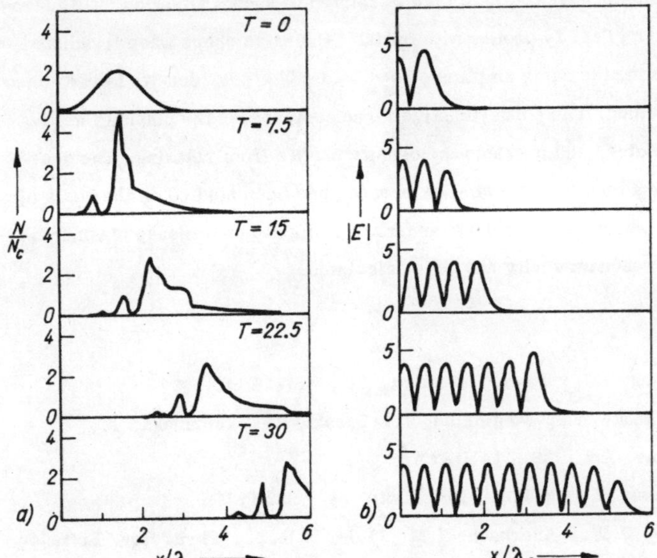

Fig. 11.14 Acceleration of a plasma with Gaussian density shape ($N_{max}/N_c = 2$, half-width $0.75\ \lambda_o$) by an electromagnetic wave with amplitude $E_o = 2.0$; spatial profiles of density (a) and field amplitude (b) for increasing times

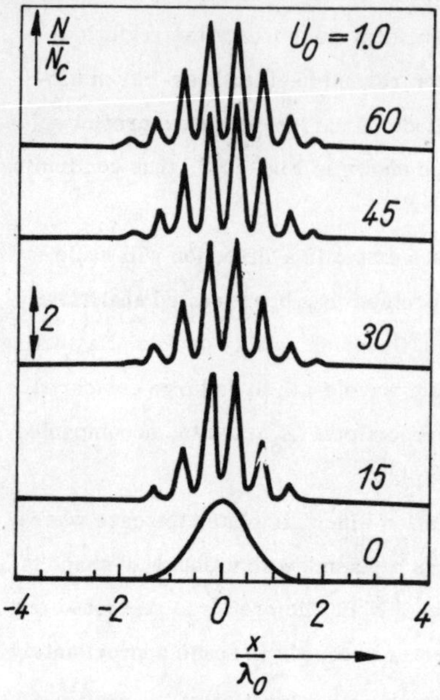

Fig. 11.15 Plasma confinement by two intense counter-propagating electromagnetic waves with amplitude $E_o = 1.0$; plasma density profiles for increasing times

The third example demonstrates the confinement of a spatially limited plasma by two counterpropagating electromagnetic waves. Fig. 11.15 is related to such a situation. It illustrates the time evolution of a plasma density profile with initial Gaussian shape when irradiated from both sides with waves having the same amplitude $E_o \simeq 1$. The peak density is two times the critical density at the beginning. The ponderomotive force compresses the plasma into the nodes of the standing wave structure and prevents the density profile from relaxing. The density peaks are maintained over a long time. A similar process has been studied on the basis of a particle simulation in /61/, where the time development of a slightly underdense plasma subject to a standing rf field in a resonant cavity has been calculated.

References

/1/ Donaldson, T.P., Spalding, I.J.; Phys. Rev. Lett. 36, 467(1976).

/2/ Azechi, H., Oda, S., Tanaka, K., Norimatsu, T., Sasaki, T., Yamanaka, T., Yamanaka, C.; Phys. Rev. Lett. 39, 144(1977).

/3/ Enright, G.D., Richardson, M.C.; Phys. Rev. Lett. 39, 932(1977).

/4/ Attwood, D.T., Sweeney, D.W., Auerbach, J.M., Lee, P.H.Y.; Phys. Rev. Lett. 40, 184(1978).

/5/ Phillion, D.W., Lerche, R.A., Rupert, V.C., Haas, R.A., Boyle, M.J.; Phys. Fluids 20, 1892(1978).

/6/ Raven, A., Willi, O.; Phys. Rev. Lett. 43, 278(1979).

/7/ Stenzel, R.L., Wong, A.Y., Kim, H.C.; Phys. Rev. Lett. 32, 654(1973).

/8/ Kim, H.C., Stenzel, R.L., Wong, A.Y.; Phys. Rev. Lett. 33, 886(1974).

/9/ Wong, A.Y., Stenzel, R.L.; Phys. Rev. Lett. 34, 727(1975).

/10/ Barinov, V.I., Gekker, I.R., Ivanov, V.A., Karfidov, D.M.; Fizika Plazmy 1, 647(1975).

/11/ Wong, A.Y.; Electromagnetic wave interaction with inhomogeneous plasmas UCLA-Report No. PPG 277(1976) and in Laser interaction and related phenomena, Vol. 4, ed. H. Schwarz, H. Hora, 1977.

/12/ DeGroot, J.S., Tull, J.E.; Phys. Fluids 18, 672(1976).

/13/ Mizuno, K., DeGroot, J.S.; Phys. Fluids 22, 2229(1979).

/14/ Probert, P., Shohet, J.L., Hatch, A.J.; IEEE Trans. Plasma Science Vol. PS-7, 9(1979).

/15/ Tanikawa, T., Wong, A.Y., Eggleston, D.L.; Phys. Fluids 27, 1416(1984).

/16/ McCall, G.H., Young, F., Ehlers, A.W., Kephart, J.F., Godwin, R.P.; Phys. Rev. Lett. 30, 1116(1973).

/17/ Gekelman, W., Stenzel, R.L.; Phys. Fluids 20, 1316(1977).

/18/ Wong, A.Y.; Nonlinear Phenomena in Laboratory and Space Plasmas Invited Paper, Int. Conf. on Plasma Physics Göteborg, 1982.

/19/ Wong, A.Y., Santoru, J., Darrow, C., Wang, L., Roederer, J.G.; Radio Science 18, 815(1983).

/20/ Gildenburg, V.B.; Zh. Eksp. Teor. Fiz. 46, 2156(1964).

/21/ Gildenburg, V.B., Fraiman, G.M.; Zh. Eksp. Teor. Fiz. 69, 1601(1975).

/22/ Gildenburg, V.B., Litvak, A.G., Fraiman, G.M.; Pisma Zh. Eksp. Teor. Fiz. 28, 433(1978).

/23/ Kull, H.; Projektgruppe Laserforschung Max-Planck-Gesellschaft, Garching Report PLF 16(1979).

/24/ Morales, G.L., Lee, Y.C.; Phys. Rev. Lett. 33, 1016(1974).

/25/ Morales, G.J., Lee, Y.C.; Phys. Fluids 20, 1135(1977).

/26/ Shukla, P.K., Spatschek, K.H.; J. Plasma Phys. 19, 387(1978).

/27/ Chen, H.H., Liu, C.S.; Phys. Rev. Lett. 39, 1147(1977).

/28/ Anderson, D., Bondeson, A., Lisak, M.; J. Plasma Phys. 22, 352(1979).

/29/ Baumgärtel, K., Sauer, K.; Phys. Lett. 70A, 107(1977).

/30/ Andreev, N.E., Silin, V.P.; Fiz. Plazmy 4, 908(1978).

/31/ Andreev, N.E., Silin, V.P., Stenchikov, G.L.; Pisma Zh. Eksp. Teor. Fiz. 28, 533(1978).

/32/ Andreev, N.E., Silin, V.P., Stenchikov, G.L.; Physica 2D, 146(1981).

/33/ Adam, J.C., Serveniere, A., Laval, G.; Phys. Fluids 25, 376(1982).

/34/ Degtyarev, L.M., Reztsov, A.S., Sagdeev, R.Z., Shapiro, V.D., Shevchenko, V.I.; Phys. Lett. 96A, 354(1983).

/35/ Giurdin Serveniere-Heron, A., Adam, J.C.; Phys. Fluids 27, 2205(1984).

/36/ Estabrook, K., Valeo, E., Kruer, W.; Phys. Lett. 49A, 109(1974).

/37/ Valeo, E., Kruer, W.; Phys. Rev. Lett. 33, 750(1974).

/38/ Biskamp, D., Welter, H.; in: Plasma Physics and Contr. Nucl. Fusion, Paper IAEA-CN-33 155-1, Wien, 1974.

/39/ Forslund, D., Kindel, J.M., Lee, K., Lindman, E.L., Morse, R.L.; Phys. Rev. A11, 679(1975).

/40/ Estabrook, K.G., Valeo, E.J., Kruer, W.; Phys. Fluids 18, 1151(1976).

/41/ Forslund, D.W., Kindel, J.M., Lee, K., Lindman, E.L.; Phys. Rev. Lett. 36, 35(1976).

/42/ Mulser, P., Kessel, C. van; Phys. Rev. Lett. 38, 902(1977).

/43/ Max, C., McKee, C.; Phys. Rev. Lett. 39, 1336(1977).

/44/ Lee, K., Forslund, D.W., Kindel, J.M., Lindman, E.L.; Phys. Fluids 20, 51(1977).

/45/ Sodha, M.S., Tripathi, V.K.; Phys. Rev. A16, 2101(1977).

/46/ Powers, L.V., Montry, G.R., Berger, R.L.; KMS Fusion, Report KMSF-U 813 (1978).

/47/ Sauer, K., Andreev, N.E., Baumgärtel, K.; Plasma Phys. and Contr. Nucl. Fusion Res. 1978, Wien, Vol. III, 187(1979).

/48/ Mulser, P., Kessel, C. van; J. Phys. D11, 1085(1978).

/49/ Mulser, P., Spindler, G.; Z. Naturforschung 34A, 1059(1979).

/50/ Sanmartin, J.R., Montanes, J.L.; J. Plasma Phys. 23, 349(1980).

/51/ Dragila, R., Limpouch, J.; Czech. J. Phys. B30, 143(1980).

/52/ Sauer, K.; Phys. Lett. 66A, 37(1978).

/53/ Virmont, J., Pellat, R., Mora, A.; Phys. Fluids 21, 567(1978).

/54/ Boris, J.P., Book, D.L.; J. Comp. Phys. 11, 38(1973).

/55/ Potter, D.; Computational Physics, John Wiley, 1973.

/56/ Rosenberg, D.U. von; Methods for the numerical solution of partial differential equations, Elsevier, New York, 1969.

/57/ Kruer, W.L., Valeo, E.J., K.G. Estabrook; Phys. Rev. Lett. 35, 1076(1975).

/58/ Andreev, N.E., Baumgärtel, K., Gradov, O.M., Sauer, K., Sünder, D.; Fiz. Plazmy 9, 254(1983).

/59/ Andreev, N.E., Gradov, O.M., Sünder, D.; Beitr. Plasmaphysik 20, 512(1980).

/60/ Gorbunov, L.M.; Usp. Fiz. Nauk 109, 631(1973).

/61/ Gitomer, S.J., Shohet, J.L.; Phys. Fluids 20, 1019(1977).

Appendix

Nonlinear three-wave coupling is a typical process occurring in nonlinear media where (a) the nonlinear dynamics can be considered as a perturbation of the linear wave solutions; (b) the lowest-order nonlinearity in the dynamic equations is quadratic in the field amplitudes and (c) the three-wave resonance conditions can be satisfied /1/. The evolution in time and one space dimension of the slowly varying amplitudes of the interacting waves are governed by the system

$$\frac{\partial E_o}{\partial t} + \sigma_o E_o + v_o \frac{\partial E_o}{\partial x} = c_o E_1 E_2 , \tag{A1}$$

$$\frac{\partial E_1}{\partial t} + \sigma_1 E_1 + v_1 \frac{\partial E_1}{\partial x} = c_1 E_o E_2^* , \tag{A2}$$

$$\frac{\partial E_2}{\partial t} + \sigma_2 E_2 + v_2 \frac{\partial E_2}{\partial x} = c_2 E_o E_1^* . \tag{A3}$$

Here E_o, E_1, E_2 are the complex amplitudes, σ_o, σ_1, σ_2 are the damping decrements and v_o, v_1, v_2 are the group velocities of the three waves having the frequencies ω_o, ω_1, ω_2 and the wave numbers k_o, k_1, k_2. c_o, c_1, c_2 are the coupling coefficients. The underlying matching conditions are

$$\omega_o = \omega_1 + \omega_2 , \tag{A4}$$

$$k_o = k_1 + k_2 . \tag{A5}$$

A lot of examples for resonant three-wave interaction from various branches of physics is listed in /1, 2/.

Three-wave interaction does not occur, for example, when the lowest-order nonlinearity in the dynamic equations is cubic. This is particularly the case in nonlinear optics of isotropic media whose second-order susceptibility tensor vanishes because of inversion symmetry.

We confine ourselves to the case where all of the three waves are of positive energy (no explosive instabilities) and where one of them is driven by an external source. This wave, denoted by the subscript o, plays the role of a pump wave that decays within the medium into two daughter waves having initially their amplitudes at the noise level. To describe the initial development of the interaction the term containing the product of the initially small waves on the right-hand side of Eq. (A1) can be ignored. Thus, we obtain a pair of linear equations describing the evolution of the decay waves under the action of a time-independent pump wave $E_o(x)$:

$$\frac{\partial E_1}{\partial t} + \sigma_1 E_1 + v_1 \frac{\partial E_1}{\partial x} = c_1 E_o E_2^* , \tag{A6}$$

$$\frac{\partial E_2}{\partial t} + \sigma_2 E_2 + v_2 \frac{\partial E_2}{\partial x} = c_2 E_o E_1^* . \tag{A7}$$

When the decay waves become comparable in amplitude to E_o, pump wave depletion sets in and the linear equations are no longer a good approximation. The subsequent evolution must be described by the full set of nonlinear equations (A 1) - (A 3). In what follows, we further simplify the analysis by assuming that the pump amplitude does not vary in space; i. e., $E_o(x) = E_o = \text{const.}$

A 1 Unlimited medium

Eqs. (A6), (A7) can simply be solved by the ansatz

$$E_1 , E_2^* \propto e^{i(Kx - \Omega t)} \tag{A8}$$

which results in the dispersion relation

$$(\Omega + i\sigma_1 - v_1 K)(\Omega + i\sigma_2 - v_2 K) + c_1 c_2^* |E_o|^2 = 0 . \tag{A9}$$

In an unbounded medium K is a free parameter. Consider a pure time-evolution at spatially homogeneous amplitudes ($K = 0$) Eq. (A9) gives, ignoring damping, the growth rate

$$\gamma^2 = + c_1 c_2^* |E_o|^2 \tag{A10}$$

($\Omega = i\gamma$). Thus, for $c_1 c_2^* > 0$ there exists an instability. In the presence of damping the growth rate is reduced and the pump amplitude has to exceed a certain threshold to excite an instability. Instead of Eq. (A10) we find

$$\gamma = -\frac{1}{2}(\sigma_1 + \sigma_2) \pm (c_1 c_2^* |E_o|^2 + \frac{1}{4}(\sigma_1 - \sigma_2)^2)^{1/2} \tag{A11}$$

and the threshold is given by

$$|E_o|_{thr} = \frac{\sigma_1 \sigma_2}{(c_1 c_2^*)^{1/2}} . \tag{A12}$$

Hence damping of only one of the decay waves is not sufficient to produce a finite threshold.

A 2 Finite spatial interaction region

For a bounded interaction region solutions of Eqs. (A6), (A7) have been studied in /3 - 5/ for the conservative case (absence of damping) and in /6 - 8/ in the presence of damping. For parallel group velocities ($v_1 v_2 > 0$) only a convective instability develops in which the decay wave amplitudes grow to steady-state values determined by the boundary values (that is, for v_1, $v_2 > 0$ by $E_1(0, t)$, $E_2(0, t)$). In the case of oppositely propagating daughter waves an absolute instability can be excited when the pump amplitude exceeds a threshold essentially determined by the length of the interaction region. The decay wave amplitudes grow without bound until saturation by pump wave depletion is reached. It should be pointed out, however, that even in a convective instability the decay wave amplitudes may become comparable to the pump amplitude. Also in such a situation pump wave depletion is responsible for the final steady-state amplitudes. Hence a convective instability may show features of an absolute one which makes a strict distinction between the two types difficult in certain cases.

Thresholds and growth rates of the absolute instability in a medium of finite extent can again be calculated using the ansatz (A8) /9/. Additional boundary conditions have to be satisfied by the decay waves. Consider an interaction region of length L and suppose that $v_1 < 0$, $v_2 > 0$. Appropriate boundary conditions are then

$$E_1(L, t) = 0 , \tag{A13}$$

$$E_2(0, t) = 0 . \tag{A14}$$

To match these conditions we modify the ansatz (A8) as follows

$$E_1(x, t) = E_1 e^{-i\Omega t} (A e^{iK_1 x} + B e^{iK_2 x}) \tag{A15}$$

with $K_{1,2}$ being solutions of the dispersion relation (A9).
$K_{1,2}$ can be represented in the form

$$K_{1,2} = U \pm W \tag{A16}$$

with

$$U = \frac{1}{2}\left\{\frac{\Omega + i\sigma_1}{v_1} + \frac{\Omega + i\sigma_2}{v_2}\right\} , \tag{A17}$$

$$W = \frac{1}{4}\left[\left\{ - \frac{\Omega + i\sigma_1}{v_1} + \frac{\Omega + i\sigma_2}{v_2}\right\}^2 - \frac{c_1 c_2^* \left|E_0\right|^2}{v_1 v_2}\right]^{1/2} . \tag{A18}$$

Condition (A13) together with ansatz (A15) yields

$$A e^{iK_1 L} + B e^{iK_2 L} = 0 . \tag{A19}$$

The boundary condition (A14) may be rewritten with the help of Eq. (6) as condition for E_1:

$$-i(\Omega + i\sigma_1) E_1(0, t) + v_1 \left. \frac{\partial E_1}{\partial x} \right/_{x=0} = 0 .$$

Using the ansatz (A15) from this equation the condition

$$-(\Omega + i\sigma_1)(A + B) + v_1 (K_1 A + K_2 B) = 0 \tag{A20}$$

can be derived. Eqs. (A19), (A20) represent a linear homogeneous system for the coefficients A and B. A non-trivial solution exists only, if its determinant vanishes. This implies

$$\left\{ -(\Omega + i\sigma_1) + v_1 K_1 \right\} e^{iK_2 L} + \left\{ (\Omega + i\sigma_1) - v_1 K_2 \right\} e^{iK_1 L} = 0 . \tag{A21}$$

Introducing Eq. (A16) gives

$$\left[-(\Omega + i\sigma_1) + v_1 (U + W) \right] e^{-iWL} + \left[(\Omega + i\sigma_1) - v_1 (U - W) \right] e^{iWL} = 0 ,$$

or

$$-i \left[U - \frac{\Omega + i\sigma_1}{v_1} \right] \sin WL + W \cos WL = 0 .$$

Multiplying this equation by L and inserting the expression (A17) for U yields the relation

$$-i \frac{L}{2} \left[-\frac{\Omega + i\sigma_1}{v_1} + \frac{\Omega + i\sigma_2}{v_2} \right] \sin WL + WL \cos WL = 0 . \tag{A22}$$

It is convenient to introduce the quantities

$$Z = \frac{L}{\pi} \left[\frac{\Omega + i\sigma_1}{|v_1|} + \frac{\Omega + i\sigma_2}{v_2} \right] , \tag{A23}$$

$$Y = \frac{2L}{\pi} \frac{\sqrt{c_1 c_2^*}\, |E_o|}{\sqrt{|v_1 v_2|}} , \tag{A24}$$

$$P = (Z^2 + Y^2)^{1/2} \tag{A25}$$

with which Eq. (A22) takes the form

$$-i \frac{Z}{P} \sin \frac{\pi}{2} P + \cos \frac{\pi}{2} P = 0 . \tag{A26}$$

This condition determines the threshold and the growth rate of the absolute instability as function of the parameters E_o, σ_1, σ_2, v_1, v_2, L. Z = 0 is seen to be a trivial solution of Eq. (A26) in the case of $\cos \pi/2 \cdot Y = 0$, i.e., for Y = 1, 3, 5, ... Nonvanishing solutions $Z \neq 0$ prove to be purely imaginary. We refer to Fig. 9.8 which presents solutions of

Eq. (A26) in the form Im Z versus Y. There is a family of solutions emerging from the zeros $Y = 1, 3, 5, \ldots$. The lowest Y attained by a solution is $Y = 1$, hence, the condition $Y = 1$ defines the threshold of the instability. In absence of damping this means

$$Y = \frac{2L}{\pi} \frac{\sqrt{c_1 c_2^*}\,|E_o|}{\sqrt{|v_1 v_2|}} \geq Y_{thr} = 1 \tag{A27}$$

A 3 The influence of a feedback

The approach developed in the preceding section is applicable also to the case of a non-zero boundary value of one of the decay waves (produced, for instance, by a feedback). A simple example of a feedback is represented by the relation

$$E_1(L, t) = R\,E_1(0, t)$$

which links the boundary values of E_1 at the edges of the interaction region. Such a feedback may turn the convective instability in the case of parallel group velocities to become absolute in nature. Moreover, a real negative feedback factor $R < 0$ gives rise to an oscillatory behaviour of the absolute instability; that is, the growth of the decay wave amplitudes is superimposed by oscillations. Similar properties exhibits the absolute instability in the case of oppositely directed group velocities when a feedback is present. Consider first the latter case. Antiparallel group velocities ($v_1 < 0$, $v_2 > 0$): We replace the boundary condition (A13) by

$$E_1(L, t) = R\,E_1(0, t) \tag{A28}$$

where R is a complex feedback factor. Using again the ansatz (A15), from this equation it follows

$$A\,e^{iK_1 L} + B\,e^{iK_2 L} = R(A + B) \tag{A29}$$

instead of Eq. (A19). The second condition (A20) remains unchanged, thus we have the modified system

$$\begin{aligned} &(e^{iK_1 L} - R)\,A + (e^{iK_2 L} - R)\,B = 0\,, \\ &\left\{-(\Omega + i\sigma_1) + v_1 K_1\right\} A + \left\{-(\Omega + i\sigma_1) + v_1 K_2\right\} B = 0 \end{aligned}$$

to determine the coefficients A, B. After the same sort of algebra as just performed, the condition for a non-zero solution to exist can be cast into the form

$$-i\frac{Z}{P}\sin\frac{\pi}{2}P + \cos\frac{\pi}{2}P = R\,e^{-i\frac{\pi}{2}Z'} \tag{A30}$$

where the additional quantity

$$Z' = \frac{L}{\pi}\left[\frac{\Omega + i\sigma_1}{v_1} + \frac{\Omega + i\sigma_2}{v_2}\right] \tag{A31}$$

has been introduced. In the limit $R \longrightarrow 0$ the condition (A30) reduces to Eq. (A26), valid in the absence of feedback. Solutions of Eq. (A30) are plotted in the form $Z = Z(Y)$ in Fig. 9.19 in the special case, that the group velocity v_1 is large compared with v_2 ($|v_1| \gg v_2$, i. e. $Z' \simeq Z$). Parallel group velocities: Consider the case v_1, $v_2 < 0$ where both decay waves propagate against the pump wave direction. Appropriate boundary conditions are then

$$E_1(L, t) = R\, E_1(0, t)\ , \tag{A32}$$

$$E_2(L, t) = 0\ . \tag{A33}$$

Condition (A32) again leads to Eq. (A29) while condition (A33) is equivalent to

$$-i(\Omega + i\sigma_1)\, E_1(L, t) + v_1 \left.\frac{\partial E_1}{\partial x}\right/_{x=L} = 0\ .$$

Using ansatz (A16) this equation can be rewritten as

$$-(\Omega + i\sigma_1)(A\, e^{iK_1L} + B\, e^{iK_2L}) + v_1(K_1 A\, e^{iK_1L} + K_2 B\, e^{iK_2L}) = 0\ .$$

Hence, the linear system for A, B has the form

$$(e^{iK_1L} - R)\, A + (e^{iK_2L} - R)\, B = 0\ ,$$

$$\left[-(\Omega + i\sigma_1) + v_1K_1\right] e^{iK_1L}\, A + \left[-(\Omega + i\sigma_1) + v_1K_2\right] e^{iK_2L}\, B = 0\ .$$

Its determinat vanishes if

$$-i\frac{L}{2}\left[\frac{\Omega + i\sigma_1}{v_1} - \frac{\Omega + i\sigma_2}{v_2}\right]\sin WL + WK\cos WL = \frac{W\,L}{R}\, e^{iUL}\ . \tag{A34}$$

This condition can be written in the form

$$i\frac{Z}{P'}\sin\frac{\pi}{2}P' + \cos\frac{\pi}{2}P' = \frac{1}{R}\, e^{i\frac{\pi}{2}Z'} \tag{A35}$$

where

$$P' = (Z^2 - Y^2)^{1/2}\ . \tag{A36}$$

Conversely to the former case, the transition to the feedback-free limit $R = 0$ is forbidden here. A finite R is necessary for an absolute instability to exist. Solutions of Eq. (A35), again in the limit $Z' \simeq Z$, are presented in Fig. 9.30 .

References

/1/ Kaup, D.J., Reimann, A., Bers, A.; Rev. Mod. Phys. 51, 275(1979).

/2/ Weiland, J., Wilhelmsson, H.; Coherent Non-linear Interaction of Waves on Plasmas. Pergamon, New York, 1977.

/3/ Manheimer, W.; Phys. Fluids 17, 1634(1976).

/4/ Harvey, R., Schmidt, G.; Phys. Fluids 18, 1385(1975).

/5/ Fuchs, V., Beaudry, G.; J. Math. Phys. 17, 208(1976).

/6/ Tang, C.L.; J. Appl. Phys. 37, 2945(1966).

/7/ Gorbunov, L.M.; Zh. Eksp. Teor. Fiz. 67, 1386(1974).

/8/ Fuchs, V.; Phys. Fluids 19, 1554(1976).

/9/ Fedorchenko, A.M., Kozarenko; Absolute and convective instability in plasmas and solids (in Russian). Nauka, Moscow, 1981.

Subject index